RETEACHING MASTERS

HOLT, RINEHART AND WINSTON

A Harcourt Classroom Education Company

Austin • **New York** • **Orlando** • **Atlanta** • **San Francisco** • **Boston** • **Dallas** • **Toronto** • **London**

To the Teacher

Algebra 1 Reteaching Masters contain two pages of alternative instruction strategies for the main skills in each lesson in *Algebra 1*. *Reteaching Masters* review key vocabulary terms and use completed examples to provide instruction for the two or three most important skills in each lesson. A set of practice exercises in which students can practice the precise skill taught follows the instruction.

Copyright © by Holt, Rinehart and Winston

All rights reserved. No part of this publication may be reproduced or transmitted in any form or by any means, electronic or mechanical, including photocopy, recording, or any information storage and retrieval system, without permission in writing from the publisher.

Teachers using ALGEBRA I may photocopy complete pages in sufficient quantities for classroom use only and not for resale.

Photo Credit
Front Cover: (background), Index Stock Photography Inc./Ron Russell; (bottom), Jean Miele MCMXCII/The Stock Market.

Printed in the United States of America

ISBN 0-03-054292-8

6   066   04   03

Table of Contents

Chapter 1	From Patterns to Algebra	1
Chapter 2	Operations in Algebra	13
Chapter 3	Equations	27
Chapter 4	Proportional Reasoning and Statistics	39
Chapter 5	Linear Functions	51
Chapter 6	Inequalities and Absolute Value	63
Chapter 7	Systems of Equations and Inequalities	73
Chapter 8	Exponents and Exponential Functions	85
Chapter 9	Polynomials and Factoring	99
Chapter 10	Quadratic Functions	115
Chapter 11	Rational Functions	127
Chapter 12	Radicals, Functions & Coordinate Geometry	139
Chapter 13	Probability	155
Chapter 14	Functions and Transformations	165
Answers		175

NAME _____ CLASS _____ DATE _____

Reteaching
1.1 Using Differences to Identify Patterns

◆ **Skill A** Extending a sequence by using differences

Recall To find the first differences, subtract each term from the next term. If the differences are not constant, use those differences and addition or subtraction to continue the sequence.

◆ **Example**
Find the next two terms in the sequence below.
2, 5, 9, 14, 20, . . .

◆ **Solution**
Find the first differences.

```
2   5   9   14   20
 \ / \ / \ / \ /
  3   4   5   6
```

Since the first differences are not constant, find the second differences.

```
2   5   9   14   20
 \ / \ / \ / \ /
  3   4   5   6
   \ / \ / \ /
    1   1   1
```

The second differences are a constant 1. Use the constant second difference to extend the sequence of first differences.

```
2   5   9   14   20
 \ / \ / \ / \ /
  3   4   5   6   7   8  →
   \ / \ / \ / \ / \ /
    1   1   1   1   1
```

2   5   9   14   20   27   35
                   \ / \ /
                    7   8

fifth term   sixth term
    ↓   ↓
  20 + **7** = 27   27 + **8** = 35
       ↑   ↑
    sixth term   seventh term

Algebra 1   Reteaching Masters   **1**

NAME _____ CLASS _____ DATE _____

Find the next two terms in each sequence using constant differences. Show your work.

1. 5, 8, 12, 17, 23, . . . _____   2. 5, 13, 21, 29, 37, . . . _____

3. 38, 29, 21, 14, . . . _____   4. 23, 22, 20, 17, 13, . . . _____

5. 8, 16, 24, 32, 40, . . . _____   6. 100, 97, 94, 91, 88, . . . _____

7. 30, 31, 35, 42, 52, . . . _____   8. 1, 4, 9, 16, 25, . . . _____

9. 12, 0, −12, −24, −36, . . . _____   10. −3, −8, −13, −18, −23, . . . _____

11. −1, 1, 7, 17, 31, . . . _____   12. 4, 3, 0, −5, −12, . . . _____

13. The fourth and fifth terms of a sequence are 39 and 54. The second differences are a constant 3. What are the first three terms of the sequence?

_____

14. The first term of a sequence is 8. The first of the first differences is 3. The second differences are a constant 1. What are the first five terms of the sequence?

_____

2   Reteaching Masters   Algebra 1

Reteaching

1.2 Variables, Expressions, and Equations

◆ **Skill A** Using a table to evaluate an expression

Recall A variable is a symbol that represents a number. An algebraic expression is a mathematical phrase that expresses a relationship between numbers, variables, and operations. Two algebraic expressions separated by an equal sign form an equation.

◆ **Example**
Noah earns $8 per hour by working at a photography studio. How much does Noah earn for working 1, 2, 3, 4, and 5 hours?

◆ **Solution**
To find Noah's earnings, make a table.

x	$8x$	y
1	8(1) = **8**	8
2	8(2) = **16**	16
3	8(3) = **24**	24
4	8(4) = **32**	32
5	8(5) = **40**	40

Noah earns $8, $16, $24, $32, and $40, respectively.

Make a table showing the value of each expression when the value of x is 1, 2, 3, 4, and 5.

1. $8x$

2. $6x + 5$

3. $26 - x$

4. $\dfrac{120}{x}$

5. $5x + 1$

6. $1 - 5x$

Algebra 1   Reteaching Masters

NAME _____ CLASS _____ DATE _____

> ◆ **Skill B**   Using guess-and-check to solve an equation
>
> **Recall**   The solution to an equation consists of a set of values for the variables that makes a true statement when substituted into the equation.
>
> ◆ **Example**
> How many $6 movie tickets can you buy with $50?
>
> ◆ **Solution**
> 1. Write an equation to model the problem.
>   Let m equal the number of tickets.
> $6m = 50$
>
> 2. Use guess-and-check to solve the equation.
>   Guess: 9   Guess: 8
> $6 \cdot 9 = 54$ $6 \cdot 8 = 48$
>   Too high   Too small
>
> You cannot buy part of a ticket, so the solution is $m = 8$.
> You can buy 8 tickets with $50.

Use guess-and-check to solve the following equations:

7. $5x + 3 = 23$ _____

8. $6x - 5 = 51$ _____

9. $12x + 16 = 100$ _____

10. $22x - 31 = 233$ _____

11. $12x + 25 = 205$ _____

12. $\frac{270}{x} + 24 = 39$ _____

Use guess-and-check to solve the following equations:

13. If 4 people decide to buy their friend a birthday gift that costs $18 and they share equally in the cost, how much should each person pay?

    _____

14. At Town Park, skaters can rent in-line skates and protective gear for $5 plus $3 per hour. If a person has $17, for how many hours can he or she skate?

    _____

4   Reteaching Masters   **Algebra 1**

NAME _____ CLASS _____ DATE _____

Reteaching
1.3 The Algebraic Order of Operations

◆ **Skill A** Evaluating numerical expressions by using the rules for the algebraic order of operations

Recall When several operations occur in the same expression,
- perform all multiplications and divisions in order from left to right and then
- perform all additions and subtractions in order from left to right.

◆ **Example 1**
Evaluate the expression $20 + 14 \div 7 \cdot 2$ by using the algebraic order of operations.

◆ **Solution**
$20 + 14 \div 7 \cdot 2 = 20 + \mathbf{2} \cdot 2$ Divide 14 by 7.
$ = 20 + \mathbf{4}$ Multiply 2 by 2.
$ = 24$ Add 20 and 4.
Thus, $20 + 14 \div 7 \cdot 2 = 24$.

◆ **Example 2**
Evaluate the expression $2 \cdot 2 - 2 \div 2$ by using the algebraic order of operations.

◆ **Solution**
$2 \cdot 2 - 2 \div 2 = \mathbf{4} - 2 \div 2$ Multiply 2 by 2.
$ = 4 - \mathbf{1}$ Divide 2 by 2.
$ = 3$ Subtract 1 from 4.
Thus, $2 \cdot 2 - 2 \div 2 = 3$.

Evaluate each expression.

1. $16 - 8 \div 4$ _____

2. $8 - 20 \div 5$ _____

3. $6 + 5 \cdot 3$ _____

4. $7 - 15 \div 5$ _____

5. $7 + 45 \div 9$ _____

6. $10 + 14 \div 7$ _____

7. $5 \cdot 9 - 4$ _____

8. $6 \cdot 3 + 2$ _____

9. $3 + 5 \cdot 2$ _____

10. $5 + 3 \cdot 4$ _____

11. $3 \cdot 6 + 4 \cdot 2$ _____

12. $36 \div 6 + 3$ _____

13. $8 + 12 \div 4 \cdot 6$ _____

14. $3 - 1 - 8 \div 4$ _____

15. $4 \cdot 2 + 10 \div 5$ _____

16. $3 \div 3 + 3 - 3 \div 3$ _____

17. $8 \cdot 2 + 4 - 1$ _____

18. $100 - 10 \cdot 5 \cdot 2$ _____

Algebra 1   Reteaching Masters   **5**

NAME _____ CLASS _____ DATE _____

◆ **Skill B** Evaluating numerical expressions that contain parentheses

Recall When several operations occur in the same expression,
- do all the evaluations inside any parentheses,
- evaluate each expression having an exponent,
- perform all multiplications and divisions in order from left to right, and then
- perform all additions and subtractions in order from left to right.

The underlined letters in the phrase below can help you remember the algebraic order of operations.

<u>P</u>lease <u>E</u>xcuse <u>M</u>y <u>D</u>ear <u>A</u>unt <u>S</u>ally

<u>P</u>arentheses
<u>E</u>xponents
<u>M</u>ultiplications
<u>D</u>ivisions
<u>A</u>dditions
<u>S</u>ubtractions

◆ **Example 1**
Simplify $(7 + 3) \div 2$.

◆ **Solution**
$(7 + 3) \div 2$
$\quad 10 \quad \div 2$ parentheses: $(7 + 3) = 10$
$\qquad\quad 5$ division: $\quad 10 \div 2 = 5$
Thus, $(7 + 3) \div 2 = 5$.

◆ **Example 2**
Simplify $4^3 + (15 - 7) \cdot 2$.

◆ **Solution**
$4^3 + (15 - 7) \cdot 2$
$4^3 + \quad 8 \quad \cdot 2$ parentheses: $\quad (15 - 7) = 8$
$64 + 8 \cdot 2$ exponents: $\quad\quad 4^3 = 64$
$64 + \quad 16$ multiplication: $\quad 8 \cdot 2 = 16$
$\quad\quad 80$ addition: $\quad\quad 64 + 16 = 80$
Thus, $4^3 + (15 - 7) \cdot 2 = 80$.

Evaluate each expression.

19. $10 \div (3 + 2)$ _____   20. $7 + (6 + 4) \cdot 2$ _____

21. $5 + (6 + 3) \cdot 2$ _____   22. $5 \cdot (2 + 3) - 7$ _____

23. $63 \div 9 + (18 - 5)$ _____   24. $7 \cdot (6 + 5) - 10$ _____

25. $2^4 \cdot (4 + 2)$ _____   26. $5^3 + 6$ _____

27. $4^3 - 30 \div 2$ _____   28. $14 + (3^3 - 7)$ _____

29. $4 \cdot 2^2$ _____   30. $(3^2 + 4) \cdot 3$ _____

6   Reteaching Masters   Algebra 1

Reteaching
1.4 Graphing With Coordinates

◆ **Skill A** Finding the coordinates for a point on the coordinate plane

Recall Coordinates give the location of a point. To locate a point (x, y) on a graph, start at the origin, $(0, 0)$. Move x units to the right or to the left along the x-axis and y units up or down parallel to the y-axis.

◆ **Example**
Give the coordinates of points A, B, C, and D.

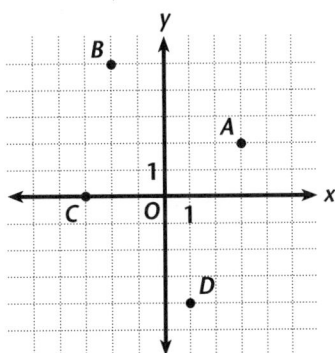

◆ **Solution**
Point A is 3 units to the right of the origin along the x-axis and 2 units up parallel to the y-axis. Thus, the coordinates of point A are $(3, 2)$.
Point B is 2 units to the left of the origin along the x-axis and 5 units up parallel to the y-axis. Thus, the coordinates of point B are $(-2, 5)$.
Point C is 3 units to the left of the origin along the x-axis and 0 units up or down parallel to the y-axis. Thus, the coordinates of point C are $(-3, 0)$.
Point D is 1 unit to the right of the origin along the x-axis and 4 units down parallel to the y-axis. The coordinates of point D are $(1, -4)$.

Give the coordinates of each point.

1. P _____
2. Q _____
3. R _____
4. S _____
5. T _____
6. U _____
7. V _____
8. W _____

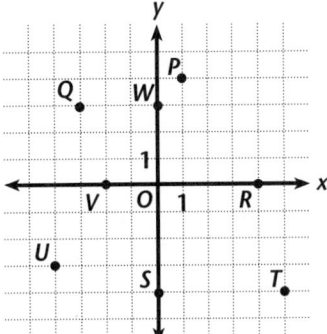

Algebra 1   Reteaching Masters   7

NAME _____ CLASS _____ DATE _____

> ◆ **Skill B** Graphing a set of points that lie along a line
>
> **Recall** If the graph of a set of ordered pairs lies along a straight line, then there is a linear equation whose graph contains those points.
>
> ◆ **Example**
> Graph $A(1, 2)$, $B(4, 4)$, and $C(8, 7)$. State whether the points lie on a straight line.
>
> ◆ **Solution**
> Graph points A, B, and C.
> If, for example, you lay a ruler so that it crosses the first and second points, A and B, it will also cross point C.
>
> If, for example, you lay a ruler so that it crosses the first and third points, A and C, it will also cross point B. The points lie on a straight line.

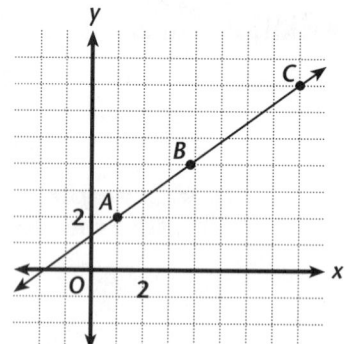

Graph each list of ordered pairs. State whether they lie on a straight line. If they do not lie on a straight line, explain why not.

9. (1, 5), (2, 7), (3, 9)

10. (2, 4), (3, 9), (4, 16)

11. (2, 2), (3, 7), (4, 12)

12. (9, 1), (12, 3), (15, 5)

Mark pays $5 for each pizza that he buys.

13. Write an equation for the total cost of the pizzas that Mark can buy. _____

14. Is the graph of the equation a straight line? Explain your response.

_____

8 Reteaching Masters   Algebra 1

NAME _____ CLASS _____ DATE _____

Reteaching
1.5 Representing Linear Patterns

◆ **Skill A** Finding the first differences and writing an equation to represent data patterns

Recall If the top row of a data table starts at 0 and increases by 1 in each column, you can find the first difference by subtracting the numbers in the second row.

◆ **Example**
Find the first differences, and write the equation that represents the pattern.

0	1	2	3	4	5	6
15	17	19	21	23	25	27

◆ **Solution**
First differences: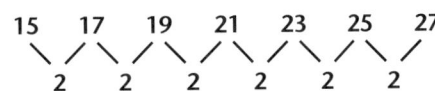

To write the equation, use the numbers in the top row for the independent variable and the numbers in the bottom row for the dependent variable. Notice that when $x = 0$, $y = 15$.

Write $y = 15 +$ (first difference) $\times x$.
The first difference is 2.
The equation is $y = 15 + 2x$.
Check another point, such as $x = 4$, in the table:
$y = 15 + 2 \cdot 4 = 15 + 8 = 23$
This y-value, 23, matches the one in the table.

Find the first differences for each data set, and write an equation to represent the data pattern.

1.
0	1	2	3	4	5	6
0	2	4	6	8	10	12

2.
0	1	2	3	4	5	6
25	21	17	13	9	5	1

3.
0	1	2	3	4	5	6
−1	−4	−7	−10	−13	−16	−19

4.
0	1	2	3	4	5	6
−25	−20	−15	−10	−5	0	5

Algebra 1 — Reteaching Masters 9

♦ **Skill B** Drawing a graph for a linear equation

Recall You can find ordered pairs for a graph by making a table.

♦ **Example**
Draw a graph of $y = 2x - 1$.

♦ **Solution**
Make a table by using 1, 2, 3, 4, and 5 for x.

1	2	3	4	5
1	3	5	7	9

The ordered pairs for the graph are (1, 1), (2, 3), (3, 5), (4, 7) and (5, 9). The horizontal axis represents x, and the vertical axis represents y.

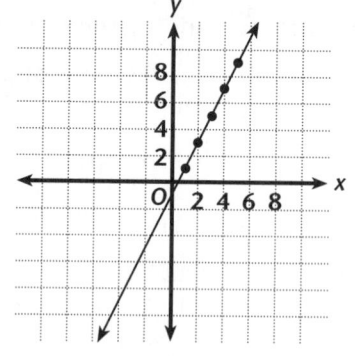

Complete the table and draw a graph for the linear equation.

5. $y = -x$

1	2	3	4	5

6. $y = 5 - 2x$

1	2	3	4	5

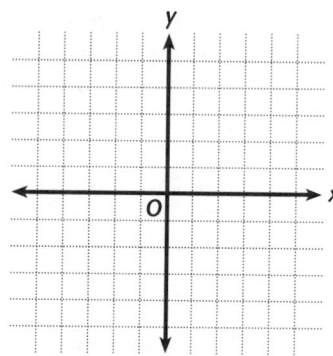

7. $y = 10 + 5x$

1	2	3	4	5

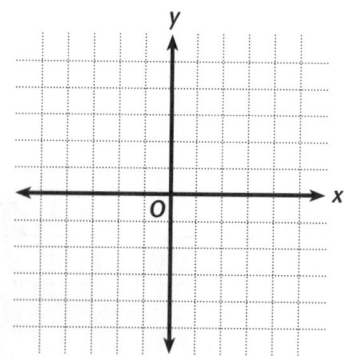

10 Reteaching Masters Algebra 1

NAME _____ CLASS _____ DATE _____

Reteaching
1.6 Scatter Plots and Lines of Best Fit

◆ **Skill A**   Drawing a scatter plot and identifying any correlation

Recall   A scatter plot is a graph of a set of ordered pairs in a coordinate plane. The ordered pairs represent data related to a pair of variables.

◆ **Example**
The information in the table shows the number of hours that cross-country runners practice per week versus their final placement in a race. Identify any correlation between the hours spent practicing and the final placement in a particular race.

Hours	3	4	5	7	8	9	10	11	12
Placement	45	34	30	20	15	17	10	9	3

◆ **Solution**
1. Draw a horizontal axis, label it "Hours per week." Draw a vertical axis, label it "Place finished." Add the appropriate numerical labels as shown at right.
2. Graph each ordered pair (hours, placement)
   The table contains nine data pairs, so your scatter plot should contain nine points.
The complete scatter plot is shown at right. Notice that the points seem to follow a down-and-to-the-right pattern, so there is a negative correlation between the variables.

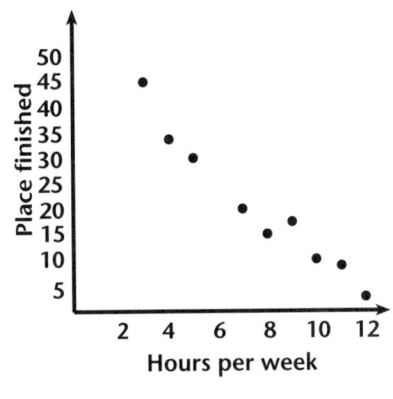

Draw a scatter plot for each set of data. Identify the correlation, if any.

1.
Miles driven	200	320	260	300	654	155	190	75	180	135
Gallons used	7.0	14.5	12.0	9.0	9.0	2.5	6.8	4.0	6.2	5.0

2. number of miners working in a given year

Year	Number of workers
1860	25,300
1880	25,000
1900	20,300
1920	18,400
1940	15,000

3. hours per week that people of a certain age spent watching television

Age	Hours
13	16
19	12
15	12
13	18
16	21
12	19
17	14
16	14

Algebra 1   Reteaching Masters   **11**

NAME _____ CLASS _____ DATE _____

◆ **Skill B**   Finding the line of best fit for a scatter plot

   Recall   The line of best fit represents an approximation of the data on a scatter plot.

   ◆ **Example**
   Approximate the line of best fit for the scatter plot.

   ◆ **Solution**
   Use a straight edge and choose the line that is closest to all the points.

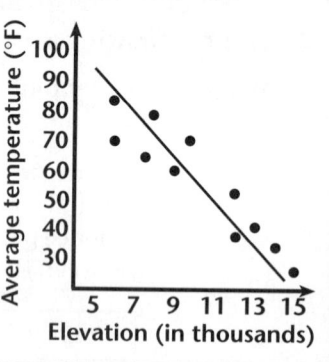

Approximate the line of best fit for each scatter plot.

4.

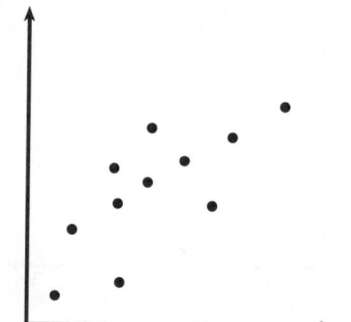

5.

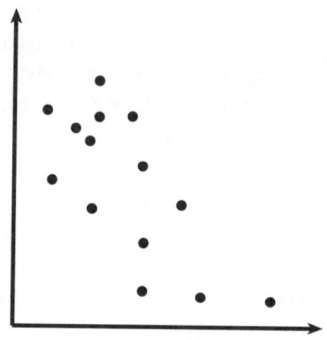

6. Which line below better represents the scatter plot at right? _____

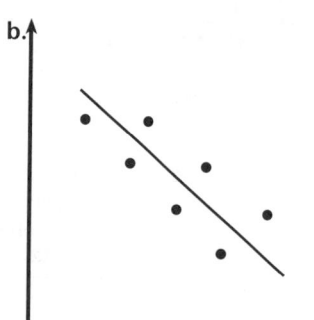

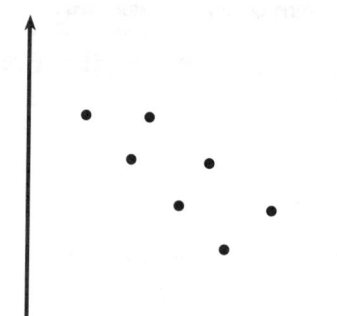

12   Reteaching Masters   Algebra 1

NAME _____ CLASS _____ DATE _____

Reteaching
2.1 The Real Numbers and Absolute Value

◆ **Skill A** Comparing real numbers

Recall The symbols used to compare two real numbers are stated and explained below.

Symbol	Meaning
<	is less than
=	is equal to
>	is greater than

◆ **Example**
Use <, >, or = to compare each pair of numbers.
a. -3 and -5 b. $-1\frac{3}{4}$ and -1.75 c. -10.2 and 8.3

◆ **Solution**
a. On a number line, -3 is located three units to the left of 0 and -5 is located five units to the left of 0, so -5 is to the left of -3 and -3 is to the right of -5 on the number line. Thus, $-3 > -5$ and $-5 < -3$.

b. Represent both $-1\frac{3}{4}$ and -1.75 as mixed numbers or as decimals. Then compare them.
When $-1\frac{3}{4}$ is represented as a decimal, it is -1.75, so $-1\frac{3}{4} = -1.75$.

c. Negative numbers are always to the left of 0 on the number line, and positive numbers are always to the right of 0 on the number line.
-10.2 is to the left of 0, and 8.3 is to the right of 0, so $-10.2 < 8.3$ and $8.3 > -10.2$.

Use <, >, or = to compare each pair of numbers.

1. 2.45 and -2.45 2. $\frac{4}{5}$ and $1\frac{4}{5}$ 3. 2.6 and $2\frac{3}{5}$ 4. -2 and $-\frac{4}{2}$

_____   _____   _____   _____

5. -11 and -12 6. $2\frac{2}{7}$ and $-\frac{16}{7}$ 7. 0 and 4.5   8. 0 and -4.5

_____   _____   _____   _____

◆ **Skill B** Finding opposites and absolute value

Recall You find the opposite of a positive number by changing its sign from $+$ to $-$.
You find the opposite of a negative number by changing its sign from $-$ to $+$.

◆ **Example**
Find the opposite of each number.
a. -10.2 b. $5\frac{1}{4}$

◆ **Solution**
a. The opposite of -10.2 is 10.2.   b. The opposite of $5\frac{1}{4}$ is $-5\frac{1}{4}$.

Algebra 1   Reteaching Masters   13

NAME _____ CLASS _____ DATE _____

Find the opposite and the absolute value of each number.

9. 2.33 _____
10. $\frac{1}{17}$ _____
11. $-\frac{9}{2}$ _____
12. $-2\frac{6}{13}$ _____

13. 12.56 _____
14. -12.56 _____
15. 1200 _____
16. -0.13 _____

17. -1356 _____
18. $3\frac{99}{100}$ _____
19. -22.7 _____
20. $100\frac{1}{2}$ _____

◆ **Skill C** Simplifying expressions involving opposites and absolute value

Recall When you simplify an expression, you follow the order of operations. You also take the simplification step by step.

◆ **Example 1**
Simplify $-(-12.8)$.

◆ **Solution**
You need to write the opposite of the opposite of 12.8. This is exactly 12.8.

◆ **Example 2**
Simplify $-|24 - 19|$.

◆ **Solution**
$-|24 - 19| = -|5|$ Perform the subtraction.
$\qquad\qquad = -5$ Take the absolute value of 5.
Thus, $-|24 - 19| = -5$.

◆ **Example 3**
Simplify $|-2| \cdot |-9|$.

◆ **Solution**
$|-2| \cdot |-9| = 2 \cdot 9$ Take the absolute value of -2 and of -9.
$\qquad\qquad = 18$ Perform the multiplication.
Thus, $|-2| \cdot |-9| = 18$.

Simplify each expression.

21. $-|-2.8|$

22. $-|2 + 9|$

23. $|-5| \cdot |6|$

24. $-(|-2| + |-2|)$

25. $|-2| - |2|$

26. $|2.5| \cdot |2.5|$

27. $|2 + 8| - |2 + 3|$

28. $|2 + 3| \cdot |2 + 3|$

29. $(2 + 6) - |6 - 2|$

14   Reteaching Masters   Algebra 1

NAME _____ CLASS _____ DATE _____

Reteaching
2.2 Adding Real Numbers

◆ **Skill A** Adding real numbers by using a number line

Recall All real numbers can be represented by points on a number line, which indicate the number's distance from 0. There is exactly one point that corresponds to each number and exactly one number that corresponds to each point.

To add two numbers on a number line, start at 0 and move to the right or left in order to reach the position of the first number. From that point, move the number of units indicated by the second number. If the second number is positive, move to the right. If the second number is negative, move to the left.

◆ **Example 1**
Add: $-4 + (-3)$.

◆ **Solution**
Locate -4 by moving 4 units to the left of 0. Then, from -4, move 3 units to the left. The ending point is at -7 on the number line.

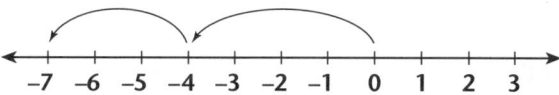

Thus, $-4 + (-3) = -7$.

◆ **Example 2**
Add: $-3 + 5$.

◆ **Solution**
Locate -3 by moving 3 units to the left of 0. Then move 5 units to the right, starting at -3. The ending point is at 2 on the number line.

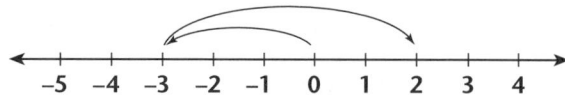

Thus, $-3 + 5 = 2$.

Use a number line to find each sum.

1. $7 + (-4)$ _____

2. $-6 + (-2)$ _____

3. $3 + (-2)$ _____

4. $-5 + (-1)$ _____

Algebra 1   Reteaching Masters   **15**

NAME _____ CLASS _____ DATE _____

◆ Skill B Adding integers

Recall The absolute value of a number gives its distance from 0 on the number line.

$$|3| = 3 \qquad |-3| = 3$$

To add integers with like signs, find the sum of the absolute values and use the sign that is common to both integers.

◆ Example 1
Add: $-6 + (-2)$.

◆ Solution
The numbers have the same sign.
Find the sum of their absolute values. $|-6| + |-2| = 6 + 2 = 8$
Write the common sign before the sum. -8

Recall To add integers with unlike signs, find the difference of the absolute values and use the sign of the integer with the greater absolute value.

◆ Example 2
Add: $4.5 + (-2.3)$.

◆ Solution
The numbers have different signs.
Subtract the smaller absolute value from
the greater absolute value. $|4.5| - |-2.3| = 4.5 - 2.3 = 2.2$
Write the sign of the number with the
greater absolute value before the sum. 2.2

◆ Example 3
Add: $\frac{1}{10} + \left(-\frac{8}{10}\right)$.

◆ Solution
$$\left|-\frac{8}{10}\right| - \left|\frac{1}{10}\right| = \frac{8}{10} - \frac{1}{10} = \frac{7}{10} \qquad \left|-\frac{8}{10}\right| > \left|\frac{1}{10}\right|$$
$$\frac{1}{10} + \left(-\frac{8}{10}\right) = -\frac{7}{10}$$

Find each sum.

5. $-3 + (-4)$ _____   6. $10 + (-11)$ _____   7. $18 + (-37)$ _____

8. $3 + (-13)$ _____   9. $-30 + (-12)$ _____   10. $-45 + 20$ _____

11. $-16 + 5$ _____   12. $-4 + 10$ _____   13. $-8 + (-14)$ _____

14. $-3.9 + (-5.6)$ _____   15. $7.8 + (-4.7)$ _____   16. $-3.6 + 12.2$ _____

17. $\frac{2}{3} + \left(-\frac{1}{3}\right)$ _____   18. $-\frac{1}{5} + \left(-\frac{3}{5}\right)$ _____   19. $-\frac{3}{8} + \left(-\frac{1}{4}\right)$ _____

Reteaching Masters — Algebra 1

Reteaching
2.3 Subtracting Real Numbers

◆ **Skill A** Subtracting real numbers by using a number line

Recall Subtraction is the inverse of addition.

To subtract one real number from another on the number line, start at 0 and move to the position of the first number. From that point, move the number of units indicated by the second number but in the opposite direction of the sign of the number.

◆ **Example**
Subtract: $3 - (-4)$.

◆ **Solution**
First, move 3 units to the right. Then, starting at 3, move 4 units to the right rather than to the left. The ending point is at 7.

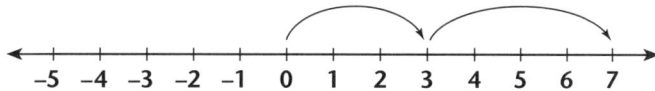

Thus, $3 - (-4) = 7$.

Find each difference. Show your work.

1. $4 - (-6)$ _____

2. $-8 - (-2)$ _____

3. $-3 - 7$ _____

4. $0 - (-3)$ _____

5. $-1 - 1$ _____

Algebra 1   Reteaching Masters   **17**

NAME _____ CLASS _____ DATE _____

◆ **Skill B** Subtracting real numbers and finding distance on a number line

Recall To subtract b from a, add the opposite of b to a.
$$a - b = a + (-b)$$

◆ **Example 1**
Find each difference.
$8 - (-3) \qquad -2.4 - 4.6 \qquad -\frac{2}{5} - \left(-\frac{1}{5}\right)$

◆ **Solution**
$8 - (-3)$ is the same as $8 + 3$. Thus, $8 - (-3) = 11$.
$-2.4 - 4.6$ is the same as $-2.4 + (-4.6)$. Thus, $-2.4 - 4.6 = -7$.
$-\frac{2}{5} - \left(-\frac{1}{5}\right)$ is the same as $-\frac{2}{5} + \frac{1}{5}$. Thus, $-\frac{2}{5} - \left(-\frac{1}{5}\right) = -\frac{3}{5}$.

Recall The distance between a and b on the number line equals $|b - a|$, or $|a - b|$.

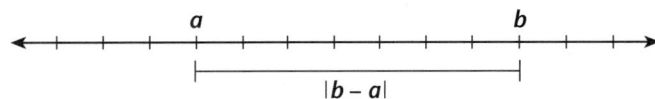

◆ **Example 2**
Find the distance between -9 and 7 on the number line.

◆ **Solution**
The distance between -9 and 7 is given by $|-9 - 7|$.
$|-9 - 7| = |-16|$
$ = 16$
Thus, the distance between -9 and 7 is 16 units.

Find each difference.

6. $-38 - (-14)$ _____
7. $28 - (-15)$ _____
8. $12 - (-28)$ _____

9. $25 - (-23)$ _____
10. $-20 - (-18)$ _____
11. $-64 - 26$ _____

12. $-1.6 - 4.5$ _____
13. $-1.4 - 1.4$ _____
14. $-2.8 - (-1.4)$ _____

15. $-3.9 - (-5.6)$ _____
16. $7.8 - (-4.7)$ _____
17. $-3.6 - 12.2$ _____

Find the distance between each pair of points on the number line.

18. 5 and 15 _____
19. -8 and -3 _____
20. -4 and 11 _____

21. -18 and -38 _____
22. 43 and -12 _____
23. 24 and 36 _____

24. -4.3 and -6 _____
25. -8.5 and -11.2 _____
26. $\frac{1}{2}$ and $-2\frac{1}{4}$ _____

18 Reteaching Masters Algebra 1

NAME _____ CLASS _____ DATE _____

Reteaching
2.4 Multiplying and Dividing Real Numbers

◆ **Skill A** Multiplying real numbers

Recall Multiplication can be thought of as repeated addition.

◆ **Example 1**
Multiply (4)(5) by using repeated addition.

◆ **Solution**
To find (4)(5), write an addition statement with 5 used four times.
$$(4)(5) = 5 + 5 + 5 + 5$$
Thus, (4)(5) = 20.

◆ **Example 2**
Multiply (4)(−5) by using repeated addition.

◆ **Solution**
To find (4)(−5), write an addition statement with −5 used four times.
$$(4)(-5) = (-5) + (-5) + (-5) + (-5)$$
Thus, (4)(−5) = −20.

Recall The product of two real numbers with the same sign is positive. The product of two real numbers with different signs is negative. Multiply the absolute values of the real numbers and apply the rule in order to get the correct sign.

◆ **Example 3**
Multiply (−4)(−5) by using the rules for multiplying signed numbers.

◆ **Solution**
Find the product of the absolute values. $|-4| \cdot |-5| = 20$
The product is positive. $(-4)(-5) = 20$
Thus, (−4)(−5) = 20.

Find each product.

1. (−2)(−7) _____

2. (8)(−11) _____

3. (12)(−7) _____

4. (−3)(−13) _____

5. (−25)(−12) _____

6. (−15)(50) _____

7. (−12)(5) _____

8. (−9)(0) _____

9. (−11)(−1) _____

10. (−1.2)(−2.5) _____

11. (3.8)(−0.1) _____

12. (−1.2)(12) _____

Algebra 1   Reteaching Masters   19

NAME _____ CLASS _____ DATE _____

◆ Skill B Dividing real numbers

Recall If the numerator and the denominator in a quotient have the same sign, the quotient is positive.
If the numerator and the denominator have different signs, the quotient is negative.

◆ **Example 1**
Find the quotient of $36 \div (-4)$.

◆ **Solution**
$36 \div 4 = 9$. The numerator and denominator in the quotient have different signs, so the quotient is negative. Thus, $36 \div -4 = -9$.

Recall To divide a real number by a real number in fraction form, multiply by the reciprocal of the divisor.

◆ **Example 2**
Find the quotient of $-8 \div \left(-\frac{1}{2}\right)$.

◆ **Solution**
The reciprocal of $-\frac{1}{2}$ is -2.

$-8 \div \left(-\frac{1}{2}\right) = -8 \cdot (-2) = 16$ *(negative · negative = positive)*

Thus, $-8 \div \left(-\frac{1}{2}\right) = 16$.

◆ **Example 3**
Find the quotient of $-\frac{2}{3} \div \frac{5}{6}$.

◆ **Solution**
The reciprocal of $\frac{5}{6}$ is $\frac{6}{5}$.

$-\frac{2}{3} \div \frac{5}{6} = -\frac{2}{3} \cdot \frac{6}{5} = -\frac{4}{5}$ *(negative · positive = negative)*

Thus, $-\frac{2}{3} \div \frac{5}{6} = -\frac{4}{5}$.

Find each quotient.

13. $(-6) \div (-2)$ _____

14. $(15) \div (-3)$ _____

15. $(12) \div (-3)$ _____

16. $(-48) \div (-4)$ _____

17. $(-75) \div \left(-\frac{3}{5}\right)$ _____

18. $(-150) \div (50)$ _____

19. $(-12.8) \div (-0.4)$ _____

20. $(0) \div (-8)$ _____

21. $(-1.8) \div (-12)$ _____

22. $(-18) \div \left(-\frac{2}{3}\right)$ _____

23. $\left(\frac{7}{8}\right) \div \left(-\frac{7}{16}\right)$ _____

24. $\left(-\frac{2}{3}\right) \div \left(\frac{3}{4}\right)$ _____

Reteaching Masters   Algebra 1

Reteaching

2.5 Properties and Mental Computation

♦ Skill A Recognizing the properties of addition and multiplication

Recall The Commutative Properties of Addition and Multiplication allow you to add or to multiply two numbers in any order.

$$a + b = b + a \qquad a \cdot b = b \cdot a$$

The Associative Properties of Addition and Multiplication allow you to regroup three numbers.

$$(a + b) + c = a + (b + c) \qquad (a \cdot b) \cdot c = a \cdot (b \cdot c)$$

The Distributive Property distributes multiplication over addition and subtraction.

$$a(b + c) = ab + ac \qquad a(b - c) = ab - ac$$

♦ Example 1
Name the property that is illustrated and complete the statement
$$38 + 43 = ? + 38$$

♦ Solution
The Commutative Property of Addition $38 + 43 = 43 + 38$

♦ Example 2
Name the property that is illustrated and complete the statement
$$30 \cdot (14 \cdot 6) = (? \cdot 14) \cdot 6$$

♦ Solution
The Associative Property of Multiplication
$30 \cdot (14 \cdot 6) = (30 \cdot 14) \cdot 6$

♦ Example 3
Use the Distributive Property to rewrite the expression $(7.5)(4.8) + (7.5)(5.2)$ as a product.

♦ Solution
$(7.5)(4.8) + (7.5)(5.2) = (7.5)(4.8 + 5.2)$

Name the property illustrated. Be specific.

1. $(23 + 4) + 19 = 23 + (4 + 19)$ _____

2. $(6 + 9) \cdot 3 = 6 \cdot 3 + 9 \cdot 3$ _____

3. $87 + 39 = 39 + 87$ _____

4. $2.5 \cdot (9 \cdot 7) = (2.5 \cdot 9) \cdot 7$ _____

5. $2.8 + 1.2 = 2.8 + 1.2$ _____

Algebra 1   Reteaching Masters

NAME _____ CLASS _____ DATE _____

Write the number that makes each statement true.

6. 27 + _____ = 12 + 27

7. _____ + 41 = 41 + 10

8. (8 + 20) + _____ = 8 + (20 + 9)

9. _____ (3.9 · 8) = (1.9 · 3.9) · 8

10. (_____)(5.2) = (5.2)(1.8)

11. 5(4.6 − 3.8) = (5)(4.6) − (_____)3.8

12. (12)(6) + (15)(6) = (12 + _____)(6)

13. 12.5 · 4.3 = _____ · (12.5)

◆ **Skill B** Using properties to mentally find products

Recall You can use properties to rearrange the numbers in a product in order to make multiplication easier.

◆ **Example**
Use properties and mental math to find the product of (25)(7)(4).

◆ **Solution**
Regroup the numbers using the Commutative Property of Multiplication and the Associative Property of Multiplication.

$$25 \cdot 7 \cdot 4 = 25 \cdot 4 \cdot 7 \quad \text{Commutative Property of Multiplication}$$
$$= (25 \cdot 4) \cdot 7 \quad \text{Associative Property of Multiplication}$$
$$= 100 \cdot 7$$
$$= 700$$

Thus, (25)(7)(4) = 700.

Use properties and mental math to find each product.

14. (5)(14.8)(2) _____

15. (2)(4.8)(50) _____

16. (20)(13)(5) _____

17. (34.6)(7.3)(0) _____

18. (4)(0.56)(25) _____

19. (2.5)(8.6)(4) _____

Use the Distributive Property and mental math to find the value of each expression.

20. (17)(4) + (17)(6) _____

21. 12(15) − 12(5) _____

22. (17)(16) − (17)(6) _____

23. 13(3) + 13(7) _____

22   Reteaching Masters   Algebra 1

Reteaching
2.6 Adding and Subtracting Expressions

◆ **Skill A** Simplifying expressions with like terms

Recall Like terms can be combined by using the Distributive Property.

◆ **Example 1**
Simplify $(-3x + 2) + (8x - 15)$.

◆ **Solution**
The terms $-3x$ and $8x$ are like terms, and 2 and -15 are like terms.
Rearrange the like terms and then simplify.

$$(-3x + 2) + (8x - 15) = (-3x + 8x) + (2 - 15)$$
$$= (-3 + 8)x + (2 - 15)$$
$$= 5x + (-13), \text{ or } 5x - 13$$

Thus, $(-3x + 2) + (8x - 15) = 5x - 13$.

◆ **Example 2**
Simplify $2(-2p + 3q) + (5p + 2q)$.

◆ **Solution**
Apply the Distributive Property, group the like terms, and then simplify.

$$2(-2p + 3q) + (5p + 2q) = 2(-2p) + 2(3q) + (5p + 2q)$$
$$= (-4p + 6q) + (5p + 2q) \quad -4p \text{ and } 5p \text{ are like terms.}$$
$$ \quad 6q \text{ and } 2q \text{ are like terms.}$$
$$= (-4p + 5p) + (6q + 2q)$$
$$= p + 8q$$

Thus, $2(-2p + 3q) + (5p + 2q) = p + 8q$.

Simplify the following expressions:

1. $(x + 6) + (3x + 3)$ _____   2. $(2m + 4) + (5m - 6)$ _____

3. $(12 - 7y) + (4y + 10)$ _____   4. $(2a + 6b) + (-6a - b)$ _____

5. $(-8x - 13) + (2x - 9)$ _____   6. $(4n + 5) + (3n + 6m)$ _____

7. $(7x - 15) + (23 - 4x)$ _____   8. $(3m + 17) + (-m - 19)$ _____

9. $(4p + 6q - 11r) + (-8p + 6q - 3r)$ _____

10. $(3x - 3y) + (12x - 2y) + (11x - y)$ _____

Algebra 1 Reteaching Masters **23**

◆ **Skill B** Subtracting expressions

Recall The opposite of $a + b$ is found by using the opposite of a and the opposite of b.
$$-(a + b) = -a + (-b) = -a - b$$

◆ **Example 1**
Simplify the expression $-(3a - 2)$.

◆ **Solution**
Write the opposite of each term.
$$-(3a - 2) = -(3a) - [-(2)]$$
$$= -3a - (-2)$$
$$= -3a + 2$$

Recall To subtract an expression, add its opposite.

◆ **Example 2**
Simplify the expression $(8a - 6b) - (10a + 2b)$.

◆ **Solution**
To subtract $10a + 2b$ from $8a - 6b$, add the opposite of $10a + 2b$.
$$(8a - 6b) - (10a + 2b) = (8a - 6b) + [-(10a + 2b)]$$
$$= (8a - 6b) + (-10a - 2b)$$
$$= (8a - 10a) + (-6b - 2b)$$
$$= -2a + -8b, \text{ or } -2a - 8b$$
Thus, $(8a - 6b) - (10a + 2b) = -2a - 8b$.

◆ **Example 3**
Simplify the expression $(2 - x) - (-2x)$.

◆ **Solution**
$$(2 - x) - (-2x) = 2 - x + 2x$$
$$= 2 + x$$
Thus, $(2 - x) - (-2x) = 2 + x$.

Simplify the following expressions:

11. $4p - 9p$ _____

12. $(8y - 17) - 9y$ _____

13. $8x - (4 + 11x)$ _____

14. $(m + 3n) - (5m + 7n)$ _____

15. $(13x - 9y) - (4x + 10y)$ _____

16. $(2a + 6b) - (14a - 11b)$ _____

17. $-(3p - 2n - t)$ _____

18. $(12x + 3y) - (6x + 2y - 4)$ _____

19. $(8a - 6b - 3) - (-2a + 4b + 7)$ _____

20. $(8m + 3n - 4) - (3m - 2n) - 8$ _____

24   Reteaching Masters   Algebra 1

NAME _____ CLASS _____ DATE _____

Reteaching
2.7 Multiplying and Dividing Expressions

◆ **Skill A** Multiplying expressions

Recall In the expression x^3, x is called the base, and 3 is called the exponent. The exponent tells how many times the base appears as a factor. For example, $x^3 = (x)(x)(x)$, $x^1 = x$, and $x^0 = 1$.

◆ **Example 1**
Simplify the expression $3x \cdot (-5x)$.

◆ **Solution**
$$3x \cdot (-5x) = (3x)(-5x)$$
$$= (3)(-5)(x)(x)$$
$$= -15x^2$$
Thus, $3x \cdot (-5x) = -15x^2$.

◆ **Example 2**
Simplify the expression $4x(2x - 3)$.

◆ **Solution**
Use the Distributive Property.
$$4x(2x - 3) = (4x)(2x) - (4x)(3)$$
$$= 8x^2 - 12x$$
Thus, $4x(2x - 3) = 8x^2 - 12x$.

◆ **Example 3**
Simplify the expression $8x^2 - 3x(x + 1)$.

◆ **Solution**
Use the definition of subtraction and then use the Distributive Property.
$$8x^2 - 3x(x + 1) = 8x^2 + (-3x)(x + 1)$$
$$= 8x^2 + (-3x)(x) + (-3x)(1)$$
$$= 8x^2 + (-3x^2) + (-3x)$$
$$= 5x^2 - 3x$$
Thus, $8x^2 - 3x(x + 1) = 5x^2 - 3x$.

Simplify the following expressions. Use the Distributive Property if needed.

1. $(-2x)(11x)$ _____

2. $5(4x^2) - 2(3x^2)$ _____

3. $-2(x^2 + x)$ _____

4. $(2x - 3x^2)6$ _____

5. $x(x + 4)$ _____

6. $6.\ 3x(x - 8)$ _____

7. $(2x + 10)(5x)$ _____

8. $-7x(5 - x)$ _____

9. $6x^2 - x(8x + 2)$ _____

10. $-3x^2 - 4x(2 - x)$ _____

Algebra 1   Reteaching Masters   **25**

◆ Skill B Dividing expressions

Recall When you divide an expression by a number, you must divide each term in the numerator by that number.

◆ **Example 1**

Simplify the expression $\frac{6a + 24}{6}$.

◆ **Solution**

Use the Distributive Property.

$\frac{6a + 24}{6} = \frac{6a}{6} + \frac{24}{6}$ Write a sum.

$\qquad\qquad = a + 4$ Divide each numerator by 6.

Thus, $\frac{6a + 24}{6} = a + 4$.

◆ **Example 2**

Simplify the expression $\frac{8(3x - 9)}{6}$.

◆ **Solution**

Use the Distributive Property.

$\frac{8(3x - 9)}{6} = \frac{24x - 72}{6}$

$\qquad\qquad = \frac{24x}{6} - \frac{72}{6}$ Write a difference.

$\qquad\qquad = 4x - 12$ Divide each numerator by 6.

Thus, $\frac{8(3x - 9)}{6} = 4x - 12$.

Simplify the following expressions. Use the Distributive Property if needed.

11. $\frac{63a}{-9}$ _____ 12. $\frac{-450n}{10}$ _____

13. $\frac{18x + 12}{-6}$ _____ 14. $\frac{8k - 12}{-4}$ _____

15. $\frac{70x - 30y}{-5}$ _____ 16. $\frac{35x + 210y}{7}$ _____

17. $\frac{3(2x + 4y)}{6}$ _____ 18. $\frac{(3x - 8)6}{2}$ _____

19. $\frac{6(5 + 10x)}{10}$ _____ 20. $\frac{12(x - 3y)}{4}$ _____

21. $\frac{(5y + 8)3.4}{2}$ _____ 22. $\frac{7(-4a - 6)}{1.4}$ _____

NAME _____ CLASS _____ DATE _____

Reteaching
3.1 Solving Equations by Adding and Subtracting

◆ **Skill A** Solving addition equations

Recall **Subtraction Property of Equality:** If equal amounts are subtracted from the expressions on each side of an equation, the expressions remain equal.

◆ **Example**
Solve $x + 80 = 230$ by using the Subtraction Property of Equality.

◆ **Solution**
$x + 80 = 230$
$x + 80 - 80 = 230 - 80$ Subtract 80 from each side of the equation.
$x = 150$

Solve each equation. Check the solution.

1. $8 + t = 64$ _____

2. $x + 19 = -25$ _____

3. $y + 4.9 = 11$ _____

4. $6 + m = -34$ _____

5. $r + 5.8 = -10.2$ _____

6. $x + 12 = 7$ _____

7. $z + \frac{3}{4} = 1\frac{3}{20}$ _____

8. $a + 15 = 12$ _____

9. $6.8 + b = -4.7$ _____

10. $x + \frac{1}{10} = \frac{1}{5}$ _____

11. The student council needs to raise $5000, but so far they have raised only $2790. Write and solve an equation to find how much money still needs to be raised.

_____

_____

12. In any triangle, the sum of the measures of the angles is 180°. Given $\triangle ABC$ with $m\angle A = 35°$ and $m\angle B = 90°$, write and solve an equation to find $m\angle C$.

_____

_____

Algebra 1   Reteaching Masters

◆ **Skill B** Solving subtraction equations

Recall **Addition Property of Equality:** If equal amounts are added to the expressions on each side of an equation, the expressions remain equal.

◆ **Example 1**
Solve the equation $y - 32 = 65$ by using the Addition Property of Equality.

◆ **Solution**
$y - 32 = 65$
$y - 32 + 32 = 65 + 32$ Add 32 to each side of the equation.
$y = 97$

◆ **Example 2**
Solve the equation $12 - x = 10$ by using the Subtraction Property of Equality.

◆ **Solution**
$12 - x = 10$
$12 - x - 12 = 10 - 12$ Subtract 12 from each side of the equation.
$-x = -2$
$x = 2$ The opposite of x is -2; therefore, $x = 2$

Solve each equation.

13. $x - 3 = 15$ _____

14. $t - 8 = -34$ _____

15. $18 = y - 7$ _____

16. $m - 2.4 = 18$ _____

17. $b - 3.8 = -13.3$ _____

18. $275 = x - 365$ _____

19. $y - 3\frac{5}{6} = 4\frac{2}{3}$ _____

20. $10 - a = 15$ _____

21. $1.8 - x = 0.2$ _____

22. $2\frac{7}{8} - y = -4\frac{1}{8}$ _____

23. $225 - b = 45$ _____

24. $17.5 - c = 28.2$ _____

25. If you spend $60 of your weekly paycheck and have $215 left, write and solve an equation to find the original amount of your weekly paycheck.

_____

28   Reteaching Masters   Algebra 1

Reteaching

3.2 Solving Equations by Multiplying and Dividing

◆ **Skill A** Solving multiplication equations

Recall **Division Property of Equality:** If each side of an equation is divided by a nonzero number, the results of both sides are equal.

◆ **Example 1**
Solve $-6x = 30$.

◆ **Solution**
$\frac{-6x}{-6} = \frac{30}{-6}$ Divide each side of the equation by –6.
$x = -5$ Simplify.

◆ **Example 2**
Solve $3w = -11$.

◆ **Solution**
$\frac{3w}{3} = \frac{-11}{3}$ Divide each side of the equation by 3.
$w = \frac{-11}{3}$, or $-3\frac{2}{3}$

Solve each equation and check your solution.

1. $-6d = -42$ _____
2. $8y = -40$ _____
3. $-7a = 49$ _____
4. $18x = -216$ _____
5. $-5s = -85$ _____
6. $3t = 64$ _____
7. $-12x = -144$ _____
8. $-5x = 22$ _____

Write and solve an equation for each situation.

9. If the perimeter of a square is 14 inches, what is the measurement of one of its sides?

10. Ricky earns $4.50 per hour working at the arcade. If he made $101.25 this week, how many hours did he work?

Algebra 1 Reteaching Masters 29

NAME _____ CLASS _____ DATE _____

◆ **Skill B** Solving division equations

Recall **Multiplication Property of Equality:** If each side of an equation is multiplied by the same number, the results of both sides are equal.

◆ **Example 1**

Solve $\frac{x}{11} = -25$.

◆ **Solution**

$(11)\frac{x}{11} = (11)(-25)$ Multiply each side of the equation by 11.

$x = -275$ Simplify.

◆ **Example 2**

Solve $\frac{y}{-3} = 9$.

◆ **Solution**

$(-3)\frac{y}{-3} = (-3)9$ Multiply each side of the equation by -3.

$y = -27$

Solve each equation and check your solution.

11. $11 = \frac{m}{5}$ _____

12. $-10 = \frac{y}{-7}$ _____

13. $\frac{a}{-4} = -16$ _____

14. $\frac{x}{2} = -34$ _____

15. $\frac{t}{13} = -24$ _____

16. $\frac{n}{-12} = 4$ _____

17. $\frac{x}{-3.6} = 14$ _____

18. $\frac{y}{-4.9} = -9.8$ _____

Write and solve an equation for each situation.

19. The slope of a line equals the change in y divided by the corresponding change in x. If the slope of a line is -6 and the change in y is 8.2, what is the change in x?

_____

_____

20. A basketball team received a care package and split it evenly among 5 players. If each player received 9 candy bars, how many candy bars were in the care package?

_____

_____

NAME _____ CLASS _____ DATE _____

Reteaching
3.3 Solving Two-Step Equations

◆ **Skill A** Solving two-step equations

Recall To solve equations, use addition, subtraction, multiplication, or division to isolate the variables.

◆ **Example 1**
Solve $3x + 2 = 17$.

◆ **Solution**

$3x + 2 = 17$
$3x + 2 - 2 = 17 - 2$ Subtract 2 from each side of the equation.
$\frac{3x}{3} = \frac{15}{3}$ Divide each side of the resulting equation by 3.
$x = 5$

◆ **Example 2**
Solve $\frac{x}{3} - 4 = 1$.

◆ **Solution**

$\frac{x}{3} - 4 = 1$
$\frac{x}{3} - 4 + 4 = 1 + 4$ Add 4 to each side of the equation.
$(3)\frac{x}{3} = 5(3)$ Multiply each side of the resulting equation by 3.
$x = 15$

Solve each equation.

1. $3x + 2 = 8$ _____
2. $2t - 4 = 8$ _____
3. $5y + 10 = 30$ _____
4. $7x + 2 = 37$ _____
5. $-2 + 7x = 33$ _____
6. $-2 + \frac{z}{2} = 1$ _____
7. $\frac{s}{4} + 2 = 6$ _____
8. $9w - 4 = 77$ _____
9. $20 + 2f = 2$ _____
10. $10 + 6c = 52$ _____
11. $8x - 5 = 43$ _____
12. $2 - \frac{t}{9} = 11$ _____
13. $32x + \frac{1}{2} = 16.5$ _____
14. $16x - \frac{1}{4} = 63.75$ _____
15. $-72 + 14j = -2$ _____
16. $-4 + 6k = -34$ _____

Algebra 1 | Reteaching Masters **31**

NAME _____ CLASS _____ DATE _____

◆ **Skill B** Write and solve equations that represent real-world situations

   Recall To solve a real-world problem, choose a variable to represent the unknown and then write an equation using that variable.

   ◆ **Example**
   The cost of going to the show includes admission plus refreshments. Suppose the admission is $7.50, a bag of popcorn costs $3.00, and 4 friends go to the show together. Find how many bags of popcorn the friends bought if the total cost for the group was $39.00.

   ◆ **Solution**
   Let x represent the number of bags of popcorn.
   Admission + Cost of popcorn = Total cost
   $$4 \cdot 7.50 + 3x = 39$$
   $$30 + 3x = 39$$
   $$30 - 30 + 3x = 39 - 30 \quad \text{Subtract 30 from each side.}$$
   $$3x = 9$$
   $$x = 3 \quad \text{Divide each side by 3.}$$
   The friends shared 3 bags of popcorn.

Write an equation and solve each problem.

17. Maria purchased 8 hats for $10 each and 6 scarves of equal value. Her total bill was $290. How much did each scarf cost?

_____

_____

_____

18. Manuel bought a set of tracks for $35 and 8 individual train cars, which were priced the same. If the total cost was $155, what was the price for each train car?

_____

_____

_____

19. The Pep Club planned to sell streamers for 50¢ each. They also raised $100 selling pom poms. How many streamers would they need to sell in order to raise a total of $150?

_____

_____

_____

32   Reteaching Masters   Algebra 1

NAME _____ CLASS _____ DATE _____

Reteaching
3.4 Solving Multistep Equations

◆ **Skill A** Solve multistep equations with variables on both sides

Recall To solve an equation with variables on both sides, you must isolate the variable.

◆ **Example**
Solve $4x - 3 = 2x + 5$.

◆ **Solution**
$4x - 3 = 2x + 5$
$4x - 3 - 2x = 2x + 5 - 2x$ Get all the terms with x on the left side of the equation.
$2x - 3 = 5$
$2x = 8$ Solve as a two-step equation.
$x = 4$

Solve and check each equation.

1. $2x + 1 = 5x - 2$ _____

2. $8y - 7 = 7y - 15$ _____

3. $4a + 2 = 8a + 18$ _____

4. $9x + 6 = 26 - x$ _____

5. $12t - 19 = 15t + 8$ _____

6. $13 - 6x = 6x + 1$ _____

7. $4y - 11 = 9 - 4y$ _____

8. $15b + 14 = 5b + 4$ _____

9. $30w - 50 = 12w - 14$ _____

10. $7p - 10 = 12 - 4p$ _____

Algebra 1   Reteaching Masters   33

NAME _____ CLASS _____ DATE _____

◆ **Skill B** Solve multi-step equations by clearing the equation of fractions

Recall The least common denominator of two fractions is the smallest number that is a multiple of the denominator of both fractions.

◆ **Example**

Solve $\frac{x}{2} + \frac{1}{3} = \frac{4x}{3} - \frac{1}{2}$

◆ **Solution**

$\frac{x}{2} + \frac{1}{3} = \frac{4x}{3} - \frac{1}{2}$ The least common denominator is 6.

$6\left(\frac{x}{2} + \frac{1}{3}\right) = 6\left(\frac{4x}{3} - \frac{1}{2}\right)$ Multiply each side of the equation by 6.

$6 \cdot \frac{x}{2} + 6 \cdot \frac{1}{3} = 6 \cdot \frac{4x}{3} - 6 \cdot \frac{1}{2}$ Distributive Property

$3x + 2 = 8x - 3$
$3x + 2 - 3x = 8x - 3 - 3x$ Isolate the variable.
$2 = 5x - 3$
$5 = 5x$ Add 3 to both sides.
$x = 1$ Divide both sides by 5.

Solve and check each equation.

11. $\frac{5x}{4} = 3 + \frac{x}{2}$ _____

12. $\frac{3}{4} - 3x = \frac{1}{4} + x$ _____

13. $\frac{1}{5} + 2n = \frac{2}{3} + 3n$ _____

14. $\frac{m}{3} = \frac{m}{2} - 1$ _____

15. $\frac{x}{8} = \frac{x}{4} + 2$ _____

16. $\frac{5y}{6} - 1 = \frac{3y}{4} + 2$ _____

17. Georgia has scored 83 and 94 on her first two history tests. What score does she need on her third test so that her average score will be exactly 90?

_____

_____

34   Reteaching Masters   Algebra 1

NAME _____ CLASS _____ DATE _____

Reteaching
3.5 Using the Distributive Property

◆ **Skill A** Solving multistep equations with variables inside parentheses

Recall Use the Distributive Property to help simplify the equation by removing the parentheses.
Distributive Property: $a(b + c) = ab + ac$ and $a(b - c) = ab - ac$

◆ **Example 1**
Solve $2(6x + 4) = 68$.

◆ **Solution**

$2(6x + 4) = 68$	Given
$2(6x) + 2(4) = 68$	Distributive Property
$12x + 8 = 68$	Simplify.
$12x = 60$	Subtraction Property of Equality
$x = 5$	Division Property of Equality

◆ **Example 2**
Solve $9t - 3(2t - 5) = 45$.

◆ **Solution**

$9t - 3(2t - 5) = 45$	Given
$9t - [3(2t) - 3(5)] = 45$	Distributive Property
$9t - 6t + 15 = 45$	Simplify.
$3t + 15 = 45$	Combine like terms.
$3t = 30$	Subtraction Property of Equality
$t = 10$	Division Property of Equality

Solve each equation.

1. $6(x + 2) = -24$ _____

2. $3(2t + 4) = 42$ _____

3. $9(z + 2) = 45$ _____

4. $8(k + 20) = 52$ _____

5. $2(x + 2) + 4 = 16$ _____

6. $6(3m - 2) = 24$ _____

7. $2h + 2(2h + 4) = 26$ _____

8. $7(2n + 3) = 91$ _____

9. $8(3t + 2) - 60 = 28$ _____

10. $-4(8c + 2) - 52 = 100$ _____

11. $4(2f - 3) + 3 = 39$ _____

12. $5(3y + 2) + 10 = 105$ _____

Algebra 1   Reteaching Masters   35

NAME _____ CLASS _____ DATE _____

◆ **Skill B**   Solving real-world problems that use equations with parentheses

Recall   To solve real-world problems, you must choose a variable and write an equation that models the situation.

◆ **Example**
The Bargain Books Store offers discounts of 50¢ per book when you buy more than 5 titles at one time. If you pay $33.75 for 5 books of equal value, what was the original cost of each book?

◆ **Solution**
Let x represent the cost of each book. The discounted price is $x - 0.50$.
5(discounted price) = 33.75
$5(x - 0.50) = 33.75$
$5x - 2.50 = 33.75$ Distributive Property
$5x = 36.25$ Add 2.50 to each side.
$x = 7.25$ Divide each side by 5.
Each book originally cost $7.25.

Write and solve an equation for each problem.

13. The CD club offers a discount when you purchase 3 or more CDs. If the price of each CD is reduced by $1.50 and the cost of 3 CDs is $35.97, find the original price of each CD.

_____

14. Write and solve an equation to determine the value of x for which the perimeter of the rectangle will equal the perimeter of the square.

_____

_____

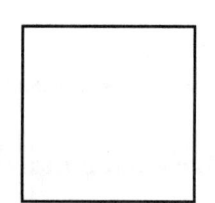

Square: side $3x$

Rectangle: width 2, length $x + 4$

15. A Chinese restaurant has the family special shown at right. What is the original average cost of the entrees if the discount is $2 per entree?

ANY FOUR ENTREES FOR $20

_____

16. Acme car rental agency charges a fee of $29 per day plus $0.15 per mile. Travel Ease car rental agency charges $20 per day plus $0.25 per mile. For a one-day trip, what mileage would make the two rates equal?

_____

36   Reteaching Masters   Algebra 1

Reteaching
3.6 Using Formulas and Literal Equations

◆ **Skill A** Rewriting a formula or a literal equation

Recall A literal equation is an equation that contains different variables. Sometimes a formula is called a literal equation when the variables represent specific quantities.

◆ **Example 1**
Given the formula $A = P + I$, write a formula for the principal, P, based on the interest, I, and amount, A.

◆ **Solution**
$A = P + I$ Given
$A - I = P + I - I$ Subtract I from each side of the equation.
$A - I = P$
$P = A - I$

◆ **Example 2**
Solve the equation $6x + 2y = 8$ for y.

◆ **Solution**
$6x + 2y = 8$ Given
$6x + 2y - 6x = 8 - 6x$ Isolate the y-term.
$2y = 8 - 6x$
$\frac{2y}{2} = \frac{8 - 6x}{2}$
$y = 4 - 3x$ Simplify.

Solve each equation for the indicated variable.

1. $x - y = 10$, for x _____

2. $x + y = z$, for y _____

3. $x - z = -y$, for z _____

4. $a + x = 2y$, for x _____

5. $x + y - z = 32$, for y _____

6. $d = rt$, for t _____

7. $A = \frac{1}{2}bh$, for h _____

8. $3x + 5y = 15$, for x _____

9. $12x - 6y = 18$, for y _____

10. $p = 2(l + w)$, for l _____

Algebra 1   **Reteaching Masters**   **37**

NAME _____ CLASS _____ DATE _____

> ◆ **Skill B** Using formulas to solve problems
>
> **Recall** When you have values for some of the quantities in a formula, you can use substitution to find the value of the remaining quantity.
>
> ◆ **Example**
> The formula for the area of a parallelogram is $A = bh$. If the area of a parallelogram is 60 square inches and the height is 10 inches, find the base.
>
> ◆ **Solution**
> $A = bh$
> $60 = b(10)$ Substitute
> $\dfrac{60}{10} = \dfrac{b(10)}{10}$ Divide both sides by 10.
> $6 = b$
> $b = 6$
>
> Thus, the base is 6 inches.

Solve each problem.

11. Use the formula for perimeter, $p = 2l + 2w$. Find w when $p = 30$ and $l = 12$.

_____

12. Use the formula for circumference of a circle, $C = 2\pi r$. Find r when $C = 14\pi$.

_____

13. The formula for the area of a triangle is $A = \dfrac{1}{2}bh$. If the area of a triangle is 75 square meters and the base has a length of 15 meters, find the height.

_____

14. The formula for distance is $d = rt$, where d is distance, r is rate, and t is time. If you travel at 80 kilometers per hour, find the amount of time that it will take to travel 320 kilometers.

_____

15. The formula for profit is $P = R - C$, where P is profit, R is revenue, and C is cost. If a company makes \$15,000 in revenue and \$8000 in profit, find the cost.

_____

16. The formula $P_1V_1 = P_2V_2$ is called Boyle's Law. P_1 and P_2 represent the pressure applied to a gas at two different times, and V_1 and V_2 represent the volume of the gas at those times. If the volume of the gas is 4 liters when the pressure is 8kPa, find the pressure when the volume is 2 liters. (kPa is the unit that measures pressure.)

_____

Reteaching Masters   **Algebra 1**

NAME _____ CLASS _____ DATE _____

Reteaching
4.1 Using Proportional Reasoning

◆ **Skill A** Writing ratios

Recall A ratio is a comparison of two quantities by division. Key words such as *to, for, per, out of, in,* and *every* may signal a ratio. You can write a ratio as a fraction.

◆ **Example**
Write a ratio to model the sentence below.
Three out of 10 freshmen take part in athletics.

◆ **Solution**
Write a fraction. Use the part, 3, as the numerator. $\frac{3}{10}$
Use the whole amount, 10, as the denominator.

Express each ratio as a fraction in lowest terms.

1. 10 to 25 _____
2. 30 for every 6 _____
3. 18 to 16 _____

4. 6 to 4 _____
5. 6 out of 17 _____
6. 2 to 15 _____

7. 8 in every 10 _____
8. 2 per 5 _____
9. 18 to 1 _____

10. 2000 out of 10,000 people surveyed opposed the bill. _____

11. In our class of 244 students, there were 8 merit scholars. _____

12. The inspector found 3 defective parts in a batch of 500. _____

◆ **Skill B** Using cross products in proportions

Recall An equation in which two ratios, $\frac{a}{b}$ and $\frac{c}{d}$, are equal is called a proportion.

In the proportion $\frac{a}{b} = \frac{c}{d}$, b and c are the means, and a and d are the extremes.

The cross products of the proportion $\frac{a}{b} = \frac{c}{d}$ are ad and bc.

In any true proportion, cross products are equal. That is, $ad = bc$.

◆ **Example**
Use cross products to find out if the proportion $\frac{3}{5} = \frac{7}{10}$ is true.

◆ **Solution**
The cross products are $3 \cdot 10 = 30$ and $5 \cdot 7 = 35$. Because $30 \neq 35$, the proportion is not true.

Algebra 1 — Reteaching Masters 39

NAME _____ CLASS _____ DATE _____

Determine whether each proportion is true. Justify your response by using cross products.

13. $\frac{4}{5} = \frac{7}{8}$ _____   14. $\frac{8}{12} = \frac{6}{9}$ _____   15. $\frac{10}{6} = \frac{20}{12}$ _____

16. $\frac{20}{100} = \frac{5}{20}$ _____   17. $\frac{14}{16} = \frac{22}{24}$ _____   18. $\frac{15}{18} = \frac{25}{36}$ _____

19. $\frac{33}{22} = \frac{24}{16}$ _____   20. $\frac{45}{108} = \frac{5}{12}$ _____   21. $\frac{36}{100} = \frac{27}{75}$ _____

◆ **Skill B** Solving proportions

Recall You can use cross products to write and solve equations.

◆ **Example 1**
Solve $\frac{10}{12} = \frac{25}{x}$.

◆ **Solution**
The cross products are $10x$ and $12 \cdot 25$, or 300.
$10x = 300$ Set the cross products equal to one another.
$x = 30$

◆ **Example 2**
Solve $\frac{x}{1.5} = \frac{3}{2}$.

◆ **Solution**
The cross products are $2x$ and $3 \cdot 1.5$, or 4.5.
$2x = 4.5$ Set the cross products equal to one another.
$x = 2.25$

Solve each proportion.

22. $\frac{6}{7} = \frac{5.4}{b}$ _____   23. $\frac{a}{5} = \frac{2}{10}$ _____   24. $\frac{18}{p} = \frac{9}{25}$ _____

25. $\frac{5}{6} = \frac{20}{r}$ _____   26. $\frac{15}{8} = \frac{k}{32}$ _____   27. $\frac{t}{2.4} = \frac{7}{8}$ _____

28. $\frac{d}{16} = \frac{3}{4}$ _____   29. $\frac{3}{1} = \frac{m}{5}$ _____   30. $\frac{2}{s} = \frac{8}{36}$ _____

31. $\frac{8}{1.6} = \frac{e}{2}$ _____   32. $\frac{180}{9} = \frac{60}{n}$ _____   33. $\frac{4}{90} = \frac{12}{z}$ _____

34. A basketball player makes 3 out of 4 free throws. Predict how many successful shots that she will make if she attempts 24 free throws in one game. _____

---

40 Reteaching Masters   Algebra 1

Reteaching
4.2 Percent Problems

◆ **Skill A** Writing percents as decimals and as fractions

Recall Percent is another name for "per 100." The symbol % is used for percent. Twenty-five percent means 25 per 100, or 25 hundredths.

◆ **Example 1**
Write each of the following percents as decimals: 25% 150% 5%

◆ **Solution**
To write a percent as a decimal, first write the number without the percent sign. Then move the decimal point two places to the left and add zeros if necessary.

$$25\% = 0.25 \qquad 150\% = 1.50, \text{ or } 1.5 \qquad 5\% = 0.05$$

◆ **Example 2**
Write 75% as a fraction in lowest terms.

◆ **Solution**
The % sign represents "per 100." Write 75 over 100 and then reduce the fraction to lowest terms.

$$75\% = \frac{75}{100} = \frac{75 \div 25}{100 \div 25} = \frac{3}{4}$$

Write each percent as a decimal.

1. 86% _____
2. 78.3% _____
3. 6% _____
4. 46% _____

5. 23% _____
6. 90% _____
7. 125% _____
8. 0.4% _____

Write each percent as a fraction in simplest terms.

9. 80% _____
10. 12.5% _____
11. 1.8% _____
12. 1% _____

13. 48% _____
14. 40% _____
15. 180% _____
16. 37.5% _____

17. Ethan is putting a 35% down payment on a house. In lowest terms, what fraction of the purchase price is his down payment?

18. Sarah was offered a pair of jeans at $\frac{1}{3}$ off or at a discount of 30%. Which option gives Sarah the better price? _____

Algebra 1 Reteaching Masters 41

◆ **Skill B** Solving percent problems

Recall Sometimes you can use a proportion to solve a percent problem.

◆ **Example 1**
What percent of 50 is 18? Use a proportion to answer the question.

◆ **Solution**
The percent is unknown, so let $\frac{x}{100}$ represent the percent.

Write and solve a proportion.
$\frac{x}{100} = \frac{18}{50}$
$50x = 1800$
$x = 36$
Thus, 36% of 50 is 18.

Recall Sometimes you can use a multiplication equation to solve a percent problem.

◆ **Example 2**
What percent of 80 is 10?
Rephrase the question as
"10 is what percent of 80?"

◆ **Solution**
10 is what percent of 80?
10 = ? percent of 80   Let a represent the numerator of the percent as a fraction.
$10 = \frac{a}{100}$ of 80
$10 = \frac{a}{100} \times 80$ Replace of with ×.
$80a = 10 \times 100$
$a = 12.5$
Thus, 10 is 12.5% of 80.

Solve each problem.

19. What is 80% of 40? _____   20. 12 is 30% of what number? _____

21. What percent of 750 is 192? _____   22. 6 is what percent of 15? _____

23. Find 62.5% of 144. _____   24. What percent of 64 is 96? _____

25. What is 85% of 150? _____   26. 55 is what percent of 88? _____

27. You buy a shirt for $30 and the sales tax is 7.5%. How much will the shirt cost after tax? _____

28. Dinner at Jake's Diner costs $60 for two people. If you wish to leave an 18% tip, what is the total cost of the dinner for two people? _____

Reteaching
4.3 Introduction to Probability

◆ **Skill A** Finding experimental probability

Recall Suppose that *t* is the number of trials in an experiment and *s* is the number of times that a successful outcome occurs. The experimental probability, *P*, of a successful outcome is given by the following ratio:

$$P = \frac{\text{number of successes}}{\text{total number of trials}} = \frac{s}{t}$$

◆ **Example 1**
Two number cubes are rolled at the same time. During 300 rolls, the sum of 4 appears 60 times. Find the experimental probability that a sum of 4 occurs during 300 rolls.

◆ **Solution**
The number of trials is 300. The number of times that a successful outcome occurs is 60. Thus, the experimental probability that a sum of 4 occurs is $\frac{60}{300} = \frac{1}{5}$, or 20%.

◆ **Example 2**
Tanner selected a chip from a jar containing red, blue, and green chips. Each time that he selected a chip, he recorded the color and returned the chip to the jar. He selected chips 80 times and recorded 20 red chips, 35 blue chips, and 25 green chips. What is the experimental probability of selecting a red chip OR a blue chip?

◆ **Solution**
The total number of trials is 80.
The number of successful red chip OR blue chip outcomes is 20 + 35 or 55. Thus, the experimental probability of selecting a red chip OR a blue chip is as follows:

$$\frac{20 + 35}{80} = \frac{55}{80} = \frac{11}{16}$$

The experimental probability is $\frac{11}{16}$, or about 69%.

Rhonda has a bag containing red, green, yellow, and blue marbles. She draws a marble from the bag and replaces it before drawing another. She has drawn 25 marbles and recorded 6 red, 8 green, 5 white, 4 yellow, and 2 blue marbles.

Find the experimental probability of each outcome.

1. She picked a red marble. _____
2. She picked a green marble. _____
3. She picked a yellow marble. _____
4. She picked a blue marble. _____
5. She picked a green marble OR a blue marble. _____
6. She did not pick a white marble. _____

Algebra 1

Reteaching Masters 43

The table shows Ray's results when he tosses a coin 80 times.

Outcome	Frequency	Total
heads	||||| ||||| ||||| ||||| ||||| ||||| ||||| ||||| |||||	45
tails	||||| ||||| ||||| ||||| ||||| ||||| |||||	35
		80

Use the results from the table above to find the experimental probability of each outcome.

7. The coin shows heads. _____   8. The coin shows tails. _____

The number of bagels sold by the Bagel Shop in an hour is recorded in the following table:

Plain	Egg	Onion	Blueberry	Sesame	Cinnamon
24	14	12	10	8	7

Use the data from the table above to find the experimental probability of each outcome.

9. A plain bagel is sold. _____   10. A sesame bagel is sold. _____

11. An onion bagel is sold. _____   12. A cinnamon bagel is sold. _____

13. A blueberry OR sesame bagel is sold. _____

14. An onion OR an egg bagel is sold. _____

Two coins are flipped at the same time for 10 tosses. The results are shown in the table.

Trial	1	2	3	4	5	6	7	8	9	10
Coin 1	H	T	H	H	T	H	T	H	H	T
Coin 2	H	H	T	H	T	T	H	T	H	H

Use the data from the table above to find the experimental probability of each outcome.

15. Both coins show the same side. _____   16. Both coins show different sides. _____

17. Both coins show heads. _____   18. Both coins show tails. _____

19. Coin 1 shows heads. _____   20. Coin 1 shows tails. _____

21. Coin 2 shows heads. _____   22. Coin 2 shows tails. _____

44   Reteaching Masters   Algebra 1

Reteaching
4.4 Measures of Central Tendency

◆ **Skill A** Finding the measures of central tendency and the range

Recall The **mean** of a set of *n* numbers is the sum of all of the numbers divided by *n*.
The **median** is the middle number of a group of numbers in ascending order.
The **mode** is the most frequently occurring number in a group of numbers.
  A group of numbers may have more than one mode.
The **range** is the difference between the highest number and the lowest number.

◆ **Example**
Find the mean, median, mode, and range of the following numbers:

$$80, 95, 82, 79, 79, 70, 80, 80$$

◆ **Solution**
The mean is the sum of all of the numbers divided by the number of values:
$$\frac{(80 + 95 + 82 + 79 + 79 + 70 + 80 + 80)}{8} = \frac{645}{8} = 80.6.$$

To find the median, write the numbers in ascending order: 70, 79, 79, 80, 80, 80, 82, and 95. There are an even number of data, so the median is the average of the two middle numbers: $\frac{(80 + 80)}{2} = 80.$

The mode is the most frequently occurring number, 80.

The range is the difference between the highest number, 95, and the lowest number, 70: $95 - 70 = 25.$

Find the mean, median, mode, and range for each set of data.

1. The daily sales of a convenience store in a week:
   $834, $1099, $765, $900, $900, $950

   mean _____ median _____ mode _____ range _____

2. ACT scores of 8 students: 18, 30, 22, 20, 28, 20, 22, 22

   mean _____ median _____ mode _____ range _____

3. Number of points scored in 8 football games: 0, 14, 3, 14, 20, 21, 28, 10

   mean _____ median _____ mode _____ range _____

4. Gallons of gas put into a car per week for the past 7 weeks:
   11, 10, 8, 7, 10, 5, 5

   mean _____ median _____ mode _____ range _____

Algebra 1   Reteaching Masters   45

NAME _____ CLASS _____ DATE _____

◆ **Skill B** Making frequency tables

Recall A frequency table has two rows, one for the data values and one for the frequency of each data value.

◆ **Example**
Use the given data to create a frequency table and to answer the questions below.

Cars sold per day: 0, 8, 5, 2, 8, 9, 3, 3, 2, 9, 7, 0, 4, 2, 4, 6, 2, 9

a. Find the mean.
b. Find the median.
c. Find the mode.

◆ **Solution**

Numbers	0	2	3	4	5	6	7	8	9
Frequency	||	||||	||	||	|	|	|	||	|||

a. Use the frequency table to add all the numbers. Then divide the total by 18.
   $\frac{83}{18} \approx 4.61$ The mean is approximately 4.61.

b. Reorder the numbers in ascending order. There are an even number of data, so find the average of the two middle terms.
   $\frac{(4+4)}{2} = 4$ The median is 4.

c. The most frequently occurring number is 2, so the mode is 2.

Create a frequency table for each set of data. Then find the mean, median, mode, and range of the data.

Test Scores		
89	75	80
89	75	70
95	100	70
89	91	95
75	75	80

5. | Scores | |
   |-----------|--|
   | Frequency | |

mean _____   median _____   mode _____   range _____

6.

Accidents by Month											
Jan.	Feb.	Mar.	Apr.	May	June	July	Aug.	Sep.	Oct.	Nov.	Dec.
3	6	1	0	2	6	2	3	3	4	1	2

Accidents	
Frequency	

mean _____   median _____   mode _____   range _____

46 Reteaching Masters   Algebra 1

NAME _____ CLASS _____ DATE _____

Reteaching
4.5 Graphing Data

◆ **Skill A** Interpreting misleading line graphs

Recall Graphs are used to display data. The scales chosen for each axis are an important part of how the graph looks.

◆ **Example**
The graphs show sales for two ice-cream shops over 5 years.
  a. How might displaying the graphs together be misleading?
  b. In what year were the sales of store A the highest?
  c. Which store has the highest sales in one year, and how much were they?

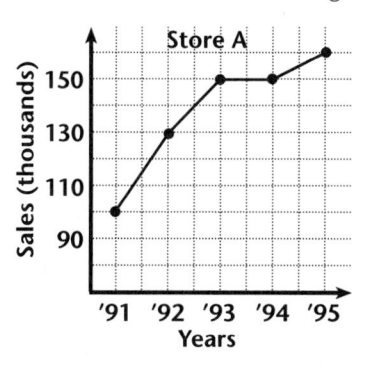

◆ **Solution**
  a. The graphs look like they use the same scale, but the graph of store B has a much higher scale, starting at $100,000 and increasing in increments of $50,000. Store B, therefore, had much higher sales.
  b. The highest point on the graph of store A is for 1995, when the store sold $160,000 worth of merchandise.
  c. Store B sold $300,000 worth of merchandise in 1995.

Use the test score graphs to answer the questions below.

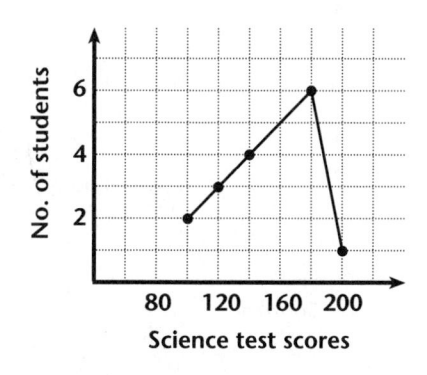

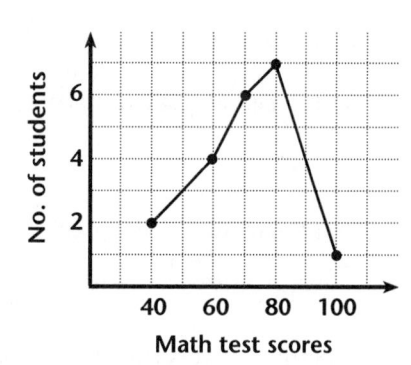

1. How might showing both graphs together be misleading?

_____

Algebra 1   Reteaching Masters 47

NAME _____ CLASS _____ DATE _____

2. What was the highest score on the science test, and how many students got that score? _____

3. What was the most frequent score on the math test? _____

4. What was the most frequent score on the science test? _____

◆ **Skill B**   Interpreting circle graphs

Recall   Circle graphs are divided into parts to show data as percents of a whole.

◆ **Example**
The graph shows the percent of students in each high-school class.

a. If there are 2580 students, how many of them are seniors?
b. Which class has 619 students?
c. Which class has the least number of students?

◆ **Solution**
a. 27% of the class are seniors.
   $0.27 \times 2580 = 697$
   There are 697 seniors.

b. $\frac{619}{2580} \approx 0.24 = 24\%$, which is the percentage for the junior class

c. The sophomore class has the least number of students.
   $23\% < 24\% < 26\% < 27\%$

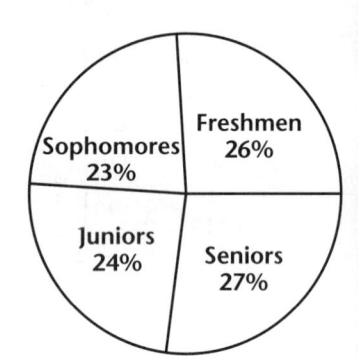

Use the circle graph shown to answer the following questions.

Favorite Types of Music
- Rock 60%
- Alternative 25%
- Jazz 10%
- Country 5%

5. If 300 people were interviewed, how many people said that rock music was their favorite? _____

6. If 600 people were interviewed, how many chose alternative music as their favorite? _____

7. If 260 people were interviewed, how many chose country as their favorite type of music? _____

8. According to this graph, what is the most popular type of music? _____

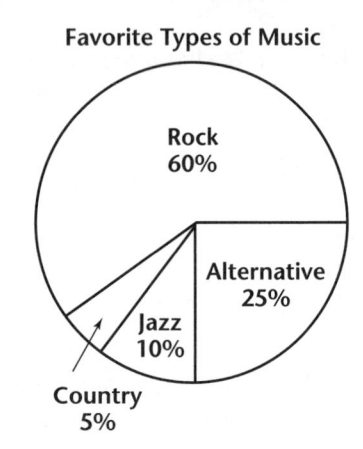

48   Reteaching Masters   Algebra 1

Reteaching
4.6 Other Data Displays

◆ **Skill A**   Creating stem-and-leaf plots

Recall   To make a stem-and-leaf plot, examine the data to decide what the stems will be. Then fill in the leaves in order from smallest to largest.

◆ **Example**
Use the data to make a stem-and-leaf plot.

   Class size: 15, 14, 15, 18, 20, 22, 20, 24, 25, 20, 32, 31, 30, 34, 35

   a. What is the range of the data?
   b. What is the mean of the data?
   c. What is the mode of the data?

◆ **Solution**
The stems will be 1, 2, and 3; fill in the leaves in order.

Stems	Leaves
1	4 5 5 8
2	0 0 0 2 4 5
3	0 1 2 4 5

a. range = 35 − 14 = 21
b. mean = $\frac{355}{15}$ = 23.7
c. mode = 20

Use the given data to make a stem-and-leaf plot and to answer each question.

Test Scores
70	65	72	70
80	82	86	80
80	86	90	92
95	90	95	94

Stems	Leaves

1. What is the range of the data? _____

2. What is the median of the data? _____

3. What is the mean of the data? _____

4. What is the mode of the data? _____

Algebra 1   Reteaching Masters   49

NAME _____ CLASS _____ DATE _____

◆ Skill B Creating box-and-whisker plots

Recall A box-and-whisker plot uses the least value, the greatest value, the median, and the upper and lower quartiles of the data.
The upper quartile is the median of all the numbers in the upper half of the data.
The lower quartile is the median of all the numbers in the lower half of the data.

◆ **Example**
Use the stem-and-leaf plot from Exercises 1–4 to make a box-and-whisker plot.

◆ **Solution**
Draw a number line and mark the least value, 65, the greatest value, 95, and the median, 84.
The lower quartile is the median of all the numbers below 84.
$$\frac{(72 + 80)}{2} = 76$$
The upper quartile is the median of the numbers above 84.
$$\frac{(90 + 92)}{2} = 91$$
Show the data on the graph and draw in the boxes, as shown.

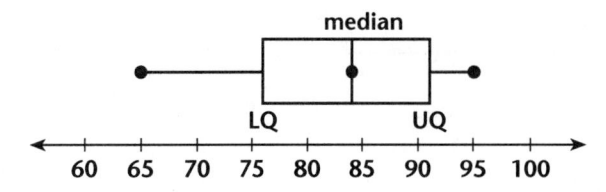

Create box-and-whisker plots for each set of data. Show your work.

5. least value _____

greatest value _____

median _____

lower quartile _____

upper quartile _____

Rainy Days per Month		
0	1	1
3	5	8
7	8	8
9	10	7
8	6	

6. Test scores for the class: 98, 76, 83, 85, 88, 94, 78, 93, 86, 86

least value _____

greatest value _____

median _____

lower quartile _____

upper quartile _____

50   Reteaching Masters   Algebra 1

NAME _____ CLASS _____ DATE _____

Reteaching
5.1 Linear Functions and Graphs

◆ **Skill A** Determining whether a relation is a function

Recall A relation is any set of ordered pairs. The first members of the ordered pairs constitute the domain of the relation, and the second members of the ordered pairs constitute the range of the relation. If each member of the domain is paired with exactly one member of the range, then the relation is a function.

◆ **Example**
Determine whether each set of ordered pairs is a function. Describe the domain and range for each.
   a. (−3.2, 5), (5, 3.2), (11.4, −3), (5.6, 7), (8, 0), (−9, 9), (5, 3.1)
   b. (−5, 5), (−4, 4), (−3, 3), (−2, 2), (0, 0), (2, 2), (3, 3), (5, 5)

◆ **Solution**
   a. To see if the relation is a function, examine the first members in the ordered pairs. If any first member is paired with more than one second member, the relation is not a function.
$$5 \longrightarrow 3.2$$
$$5 \longrightarrow 3.1$$
Because 5 is paired with both 3.2 and 3.1, the relation is not a function.
domain: {−3.2, 5, 11.4, 5.6, 8, −9}
range: {5, 3.2, −3, 7, 0, 9, 3.1}

   b. For each first member, there is a different second member, so this relation is a function.
domain: {−5, −4, −3, −2, 0, 2, 3, 5}
range: {5, 4, 3, 2, 0}

Determine whether each set of ordered pairs is a function.
Describe the domain and range for each.

1. {(1.3, −1.3), (3.1, 2.3), (10, 10), (12, 21)}

   _____

   domain: _____

   range: _____

2. {(1, 64), (2, 32), (3, 16), (4, 8), (5, 4), (6, 2)}

   _____

   domain: _____

   range: _____

3. {(−3, 7), (−5, 7), (8, 7), (10, 7), (−3, 8)}

   _____

   domain: _____

   range: _____

4. {(18, 11), (20, 12), (−20, −8), (−7, −1.5), (8, −1.5)}

   _____

   domain: _____

   range: _____

Algebra 1   Reteaching Masters

NAME _____ CLASS _____ DATE _____

◆ **Skill B** Determining a range value or a domain value given one of them and an equation

Recall When you solve an equation in one variable, you use the properties of equality.

◆ **Example**
The variables x and y are related by the equation $4x + 7y = 28$. Find each domain or range value. Then write the ordered pair that satisfies the equation.
  a. Find x given that $y = 3$. **b.** Find y given that $x = -2$.

◆ **Solution**
  a. Substitute 3 for y in $4x + 7y = 28$. Then solve for x.
  $$4x + 7(3) = 28 \qquad \text{Let } y \text{ equal 3.}$$
  $$4x + 21 = 28$$
  $$4x = 7 \qquad \text{Subtract 21 from each side of the equation.}$$
  $$x = \frac{7}{4}, \text{ or } 1\frac{3}{4} \qquad \text{Divide each side by 4.}$$
  Thus, $\left(3, 1\frac{3}{4}\right)$ satisfies $4x + 7y = 28$.

  b. Substitute -2 for x in $4x + 7y = 28$. Then solve for y.
  $$4(-2) + 7y = 28 \qquad \text{Let } x \text{ equal } -2.$$
  $$-8 + 7y = 28$$
  $$7y = 36 \qquad \text{Add 8 to each side of the equation.}$$
  $$y = \frac{36}{7}, \text{ or } 5\frac{1}{7} \qquad \text{Divide each side by 7.}$$
  Thus, $\left(-2, 5\frac{1}{7}\right)$ satisfies $4x + 7y = 28$.

Complete each ordered pair so that it is a solution to the given equation.

5. $(-4, ?)$ and $(?, 10)$; $2x + \frac{1}{2}y = 8$

6. $(0, ?)$ and $\left(?, \frac{1}{4}\right)$; $-\frac{1}{2}x - 3y = 15$

◆ **Skill C** Writing a linear equation to represent a table of values

Recall A relationship is linear if y changes by a constant amount for a fixed increase in x-values.

◆ **Example**
Write a linear equation in x and y for this table.

x	1	2	3	4	5
y	7	10	13	16	19

◆ **Solution**
For each increase of 1 in the x-values, y increases by 3. Thus, when $x = 0$, $y = 7 - 3 = 4$. An equation that represents the table is $y = 3x + 4$.

Write a linear equation to represent each table of values.

7.

x	1	2	3	4	5
y	5	11	17	23	27

8.

x	1	2	3	4	5
y	4.4	8.4	12.4	16.4	20.4

NAME _____ CLASS _____ DATE _____

Reteaching
5.2 Defining Slope

◆ **Skill A** Finding the slope of a line from its graph

Recall Slope is the ratio of vertical change to horizontal change between two points on a line. A line that slants upward from left to right has a positive slope. A line that slants downward from left to right has a negative slope. A horizontal line has a slope of zero. The slope of a vertical line is undefined.

◆ **Example 1**
Tell whether the slope of the line is positive or negative.

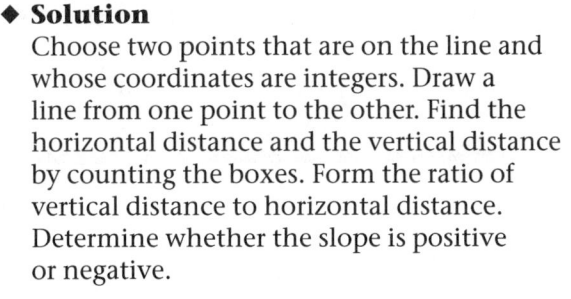

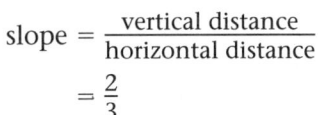

◆ **Solution**
Because the line slants downward from left to right, the slope is negative.

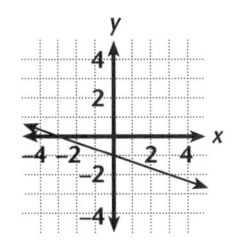

◆ **Example 2**
Find the slope of the line shown at the right.

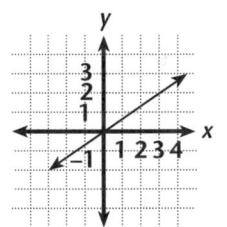

◆ **Solution**
Choose two points that are on the line and whose coordinates are integers. Draw a line from one point to the other. Find the horizontal distance and the vertical distance by counting the boxes. Form the ratio of vertical distance to horizontal distance. Determine whether the slope is positive or negative.

$$\text{slope} = \frac{\text{vertical distance}}{\text{horizontal distance}}$$

$$= \frac{2}{3}$$

The line slants upward from left to right, so the slope is $+\frac{2}{3}$.

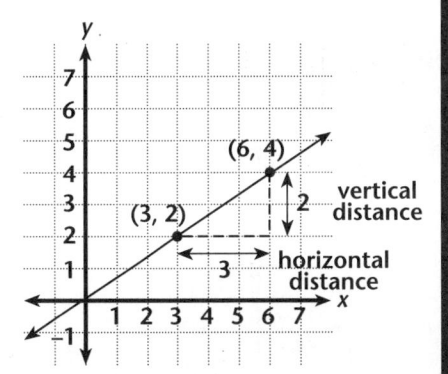

Tell whether the slope of each line is positive, negative, zero, or undefined.

1.

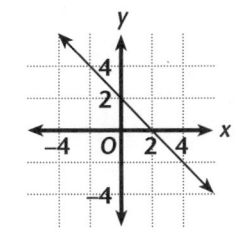

2.

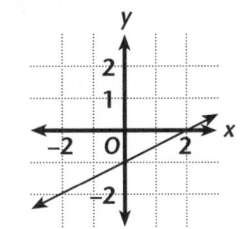

3.

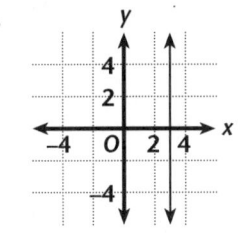

_____   _____   _____

Algebra 1   Reteaching Masters   53

NAME _____ CLASS _____ DATE _____

◆ **Skill B** Calculating the slope of a line

Recall slope $= \dfrac{\text{rise}}{\text{run}} = \dfrac{\text{difference in } y\text{-coordinates}}{\text{difference in } x\text{-coordinates}}$

◆ **Example 1**
Find the slope of the line by using the graph.

◆ **Solution**
1. Locate two points whose coordinates are on the line and are easy to read.
   (0, 3) and (1, 1)

2. Find the rise and the run as you go from one point to the other. The starting point does not matter.

slope $= \dfrac{\text{rise}}{\text{run}} = \dfrac{1-3}{1-0} = \dfrac{-2}{1} = -2$

slope $= \dfrac{\text{rise}}{\text{run}} = \dfrac{3-1}{0-1} = \dfrac{2}{-1} = -2$

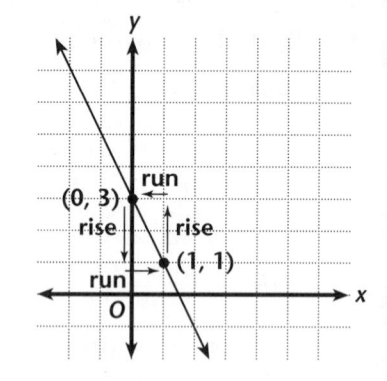

◆ **Example 2**
If the points $A(-1, 4)$ and $B(3, -2)$ are on a line, what is the slope of the line?

◆ **Solution**
slope $= \dfrac{\text{difference in } y\text{-coordinates}}{\text{difference in } x\text{-coordinates}} = \dfrac{4-(-2)}{-1-3} = \dfrac{4+2}{-4} = -\dfrac{6}{4} = -\dfrac{3}{2}$

Find the slope of each line by using the graph.

4.

5.

6.

_____   _____   _____

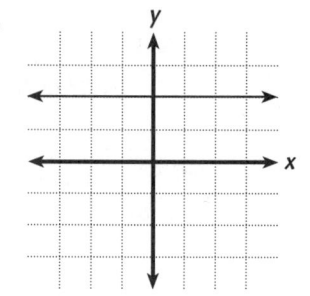

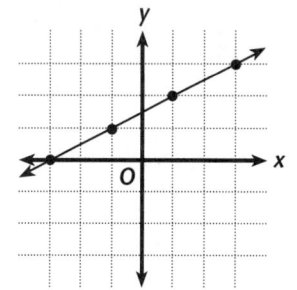

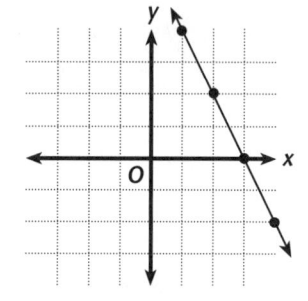

Find the slope of the line that contains each pair of points.

7. (1, 1), (4, 4) _____ **8.** (0, 1), (1, 0) _____ **9.** (−3, −1), (0, 5) _____

10. (2, 3), (−2, 1) _____ **11.** (−3, −2), (2, 1) _____ **12.** (−3, 3), (4, −1) _____

54   Reteaching Masters   Algebra 1

NAME _____ CLASS _____ DATE _____

Reteaching
5.3 Rate of Change and Direct Variation

◆ **Skill A** Finding the rate of change of a linear function by using a graph

Recall The rate of change of a linear function is the same as the slope of the graph of a linear function.

◆ **Example 1**
Find the rate of change in the linear function graphed at right.

◆ **Solution**
The rate of change or slope is

$$\frac{\text{change in the } y\text{-coordinates}}{\text{corresponding change in the } x\text{-coordinates}}$$

rate of change: $= \frac{5-3}{4-2} = \frac{2}{2} = 1$

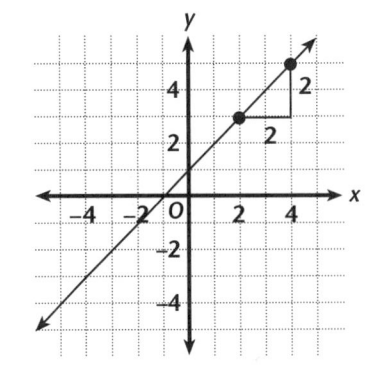

Recall When you know the coordinates of two points on a line, the slope of the graph of the linear function is given as follows:

$$\frac{\text{difference in the } y\text{-coordinates}}{\text{corresponding difference in the } x\text{-coordinates}}$$

◆ **Example 2**
Find the rate of change in the linear function whose graph contains (3, 1) and (5, 2).

◆ **Solution**
rate of change: $\frac{2-1}{5-3} = \frac{1}{2}$

Find the rate of change of the linear function containing each pair of points.

1. (5, 3) and (8, 7) _____

2. (0, 5) and (−3, 4) _____

3. (4, 9) and (2, 5) _____

4. (−6, −6) and (4, 2) _____

5. (8, 6) and (−3, 6) _____

6. (7, 2) and (2, −7) _____

7. (−3, −8) and (−1, −5) _____

8. (9, −4) and (9, 4) _____

9. Write the expression *8 out of 10 students study at least 2 hours a day* as a rate of change. _____

Algebra 1   Reteaching Masters   **55**

NAME _____ CLASS _____ DATE _____

♦ **Skill B** Solving direct-variation problems by using a table

Recall In the equation $y = mx$, m is the slope of the graph of the equation.
When $y = mx$ and $m \neq 0$, then y varies directly as x.
The value of m is called the constant of variation.
If $m > 0$, then as x increases, y increases.
If $m < 0$, then as x increases, y decreases.

♦ **Example 1**
Suppose that y varies directly as x and $y = 3x$.
a. Find the constant of variation.
b. What is the value of y when $x = 9$?

♦ **Solution**
a. The slope of $y = 3x$ is 3, so the constant of variation is 3.
b. When $x = 9$, $y = 3(9)$, or 27.

♦ **Example 2**
Jill works as a cable technician and charges by the hour. Her records show the hours worked and the salary that she receives for various jobs.

Hours	3	5	7	10	15
Salary	$135	$225	$315	$450	$675

Find the constant of variation and write the direct-variation equation.

♦ **Solution**
The ratio of Jill's salary to hours worked is constant.
$$\frac{135}{3} = \frac{225}{5} = \frac{315}{7} = \frac{450}{10} = \frac{675}{15}$$
Let x represent hours worked, and let y represent dollars earned.
Then $\frac{y}{x} = \frac{135}{3}$, or $\frac{y}{x} = 45$. Thus, 45 is the constant of variation, and $y = 45x$.

For each direct variation, find the constant of variation and an equation. Then complete the table.

10.

x	2	5	9	12	15
y	$13	$32.50	$58.50		

11.

x		56	77	112	140
y	3	8	11	16	

56 Reteaching Masters **Algebra 1**

NAME _____ CLASS _____ DATE _____

Reteaching
5.4 The Slope-Intercept Form

◆ **Skill A** Writing an equation of a line in slope-intercept form

Recall The slope-intercept form of a line is $y = mx + b$.
            ↑   ↑
         slope y-intercept

◆ **Example**
Write an equation for each line.
a. containing (0, 1) and with a slope of −2
b. containing (3, −4) and (9, 0)

◆ **Solution**
a. The slope, m, is given as −2. The line contains (0, 1), so this point is the y-intercept, or b is 1. Substituting these numbers into the equation gives $y = -2x + 1$.

b. First find the slope. $m = \frac{-4 - 0}{3 - 9} = \frac{-4}{-6} = \frac{2}{3}$

Then subtitute the coordinates of one of the given points into the equation and solve for b.

For the point (9, 0): $0 = \frac{2}{3}(9) + b$
$0 = 6 + b$
$b = -6$

Substituting this number for b and $\frac{2}{3}$ for m into the equation $y = mx + b$ gives the equation $y = \frac{2}{3}x - 6$.

For each equation, find the slope and the y-intercept.

1. $y = 3x - 1$ _____   2. $y = \frac{1}{2}x + 2$ _____   3. $y = -x + \frac{1}{2}$ _____

Write an equation in slope-intercept form for each line.

4. with a slope of 2 and a y-intercept of −1 _____

5. containing (0, −3) and with a slope of $\frac{1}{3}$ _____

Write an equation in slope-intercept form for the line that contains each pair of points.

6. (1, 1) and (3, 5) _____   7. (2, −4) and (−1, 5) _____

8. (2, 4) and (−4, 1) _____   9. (1, 0) and (3, 2) _____

Algebra 1   Reteaching Masters **57**

NAME _____ CLASS _____ DATE _____

◆ **Skill B**   Using the slope-intercept form to write equations of lines

Recall   In the formula $y = mx + b$, m is the slope and b is the y-intercept.

$$\text{slope} = \frac{\text{rise}}{\text{run}} = \frac{\text{difference in } y\text{-values}}{\text{difference in } x\text{-values}}$$

The y-intercept is the point where the line crosses the y-axis.

◆ **Example**
Find the equation for the graph.

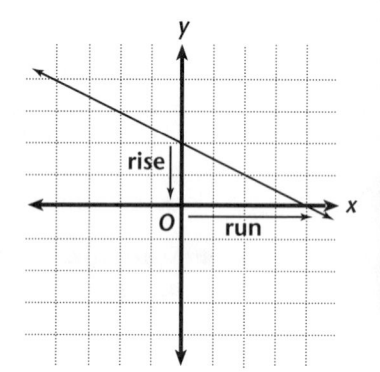

◆ **Solution**
The line crosses the y-axis at (0, 2), so $b = 2$.

Locate two points on the line and count in order to find the slope.

$$\text{slope} = \frac{\text{rise}}{\text{run}} = -\frac{2}{4} = -\frac{1}{2}$$

The equation for the graph is $y = -\frac{1}{2}x + 2$.

Write an equation in slope-intercept form for each line graphed below.

10.

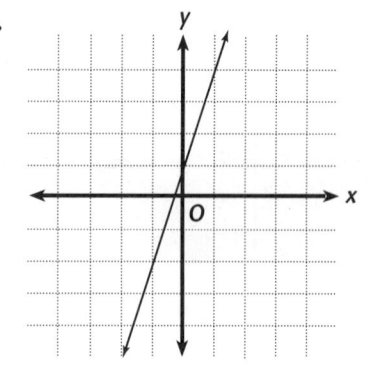

11.

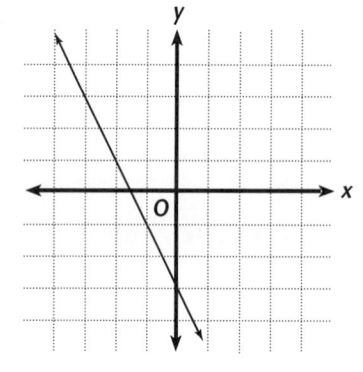

12.

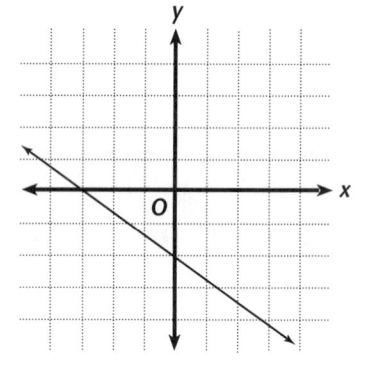

Adena charges $5 plus $3 an hour for babysitting.

13. Write an equation for the amount of money, y, that Adena can earn by babysitting for x hours. _____

14. How much will Adena earn if she babysits for 5 hours? _____

15. How many hours will Adena have to babysit in order to earn $26? _____

58   Reteaching Masters   Algebra 1

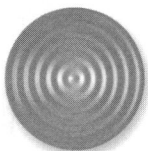

Reteaching
5.5 The Standard and Point-Slope Forms

◆ **Skill A** Writing an equation of a line in standard form

Recall The standard form for the equation of a line is $Ax + By = C$, where A, B, and C are integers, A and B are not both zero, and A is not negative.

◆ **Example**
Write the equation $x = \frac{3}{4}y + 3$ in standard form. Then find the intercepts and use them to graph the equation.

◆ **Solution**

$x = \frac{3}{4}y + 3$ Given
$4x = 3y + 12$ Multiply each side by 4.
$4x - 3y = 12$ Subtract $3y$ from each side.

To find the y-intercept, let $x = 0$.
$4x - 3y = 12$
$0 - 3y = 12$
$y = -4$

To find the x-intercept, let $y = 0$.
$4x - 3y = 12$
$4x - 0 = 12$
$x = 3$

Thus, the intercepts are $(0, -4)$ and $(3, 0)$.

Use your intercepts to graph the line.

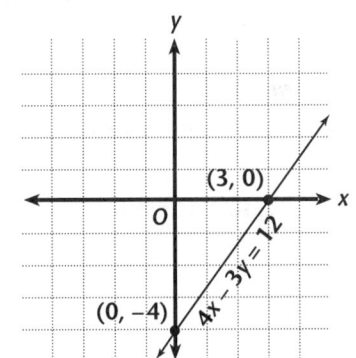

Write each equation in standard form.

1. $5x = 10y + 15$ _____
2. $3y = 2x$ _____
3. $y = \frac{2}{5}x - 1$ _____
4. $3x - 5 = \frac{1}{2}y + 1$ _____

Find the x- and y-intercepts for the graph of each equation. Then graph both equations on the grid provided.

5. $2x + y = -2$ _____

6. $3x - 2y = 6$ _____

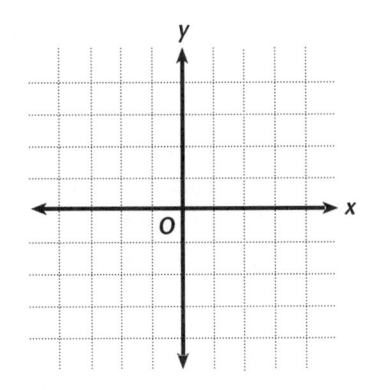

Algebra 1   Reteaching Masters **59**

NAME _____ CLASS _____ DATE _____

◆ **Skill B** Writing an equation of a line in point-slope form

Recall The point-slope form for an equation of a line is $y - y_1 = m(x - x_1)$.

◆ **Example**
Write an equation for the line through $(1, -1)$ and $(-1, 5)$
  a. in point-slope form.
  b. in slope-intercept form.

◆ **Solution**
  a. First find m.
  $$m = \frac{\text{difference in } y\text{-values}}{\text{difference in } x\text{-values}} = \frac{-1 - 5}{1 - (-1)} = \frac{-6}{2} = -3$$

  Substitute the slope and one of the points into the point-slope equation.
  $y - y_1 = m(x - x_1)$
  $y - (-1) = -3(x - 1)$ Use the point $(1, -1)$.
  $y + 1 = -3(x - 1)$ Simplify.

  b. Rewrite the equation in the form $y = mx + b$.
  $y + 1 = -3(x - 1)$
  $y + 1 = -3x + 3$ Distributive Property
  $y = -3x + 2$ Subtract 1 from each side.

Write an equation for each line in point-slope form.

7. containing $(4, -1)$ and with a slope of $\frac{1}{2}$ _____

8. crossing the x-axis at $x = -3$ and the y-axis at $y = 6$ _____

9. containing the points $(-6, -1)$ and $(3, 2)$ _____

Rewrite each equation in slope-intercept form.

10. the line from Exercise 7 _____

11. the line from Exercise 8 _____

12. the line from Exercise 9 _____

13. In what situations would you find it easier to use point-slope form, and in what situations would you find it easier to use slope-intercept form? _____

_____

_____

Reteaching
5.6 Parallel and Perpendicular Lines

◆ **Skill A** Writing an equation of a line that is parallel to a given line

Recall If two different lines have the same slope, the lines are parallel.
If two different lines are parallel, they have the same slope.
All vertical lines and all horizontal lines are parallel.

◆ **Example**
Write an equation for a line that contains the point (3, 5) and that is parallel to $2x - y = 3$.

◆ **Solution**
Step 1 To find the slope of the given equation, write an equation in the form $y = mx + b$.

$2x - y = 3$
$-y = -2x + 3$ Subtract $2x$ from each side.
$y = 2x + 3$ Multiply each side by -1.

The slope is 2. The slope of a parallel line must also be 2.

Step 2 Use the point-slope form for the equation of a line.

$y - y_1 = m(x - x_1)$
$y - 5 = 2(x - 3)$ Substitute the slope and the point (3, 5) into the equation.

Find the slope of a line that is parallel to the following lines:

1. $y = -\frac{1}{2}x + 1$ _____   2. $3x + y = 5$ _____

3. $12 = 2x - 3y$ _____   4. $x + \frac{1}{4}y = 1$ _____

Write an equation in point-slope form for each line according to the given information. Show your work.

5. containing $(-1, -4)$ and parallel to $y = 3x + 2$ _____

6. containing $(2, -4)$ and parallel to $x - 2y = 5$ _____

7. containing $(-2, 3)$ and parallel to $x = 1$ _____

8. containing $(4, 15)$ and parallel to $-x + \frac{2}{3}y = 6$ _____

9. containing $(-1, -6)$ and parallel to $y = -1$ _____

Algebra 1   Reteaching Masters **61**

NAME _____ CLASS _____ DATE _____

◆ **Skill B** Writing an equation of a line that is perpendicular to a given line

Recall If the slopes of two lines are m and $-\frac{1}{m}$ and $m \neq 0$, the lines are perpendicular.

If two lines are perpendicular and neither line is vertical or horizontal, the slopes of the lines are m and $-\frac{1}{m}$.

A line perpendicular to a horizontal line is a vertical line with an undefined slope. A line perpendicular to a vertical line is a horizontal line with a slope of 0.

◆ **Example**
Write an equation for the line that contains $(2, -5)$ and that is perpendicular to the graph of $3x + y = 1$.

◆ **Solution**
Step 1 To find the slope of the given line, write the equation in the form $y = mx + b$.

$$3x + y = 1$$
$$y = -3x + 1$$

The slope is -3, so a line perpendicular to the given line has a slope of $\frac{1}{3}$.

Step 2 Use the point-slope form for the equation of a line.

$$y - y_1 = m(x - x_1)$$
$$y - (-5) = \frac{1}{3}(x - 2) \qquad \text{Substitute the slope and the point } (2, -5).$$
$$y + 5 = \frac{1}{3}(x - 2)$$

Find the slope of a line that is perpendicular to the following lines. If the slope of the desired line is undefined, write *undefined*.

10. $y = 2x - 3$ _____

11. $\frac{1}{3}x - y = 1$ _____

12. $12 = 3x + 4y$ _____

13. $3x + 6y = 15$ _____

14. $y = -4$ _____

15. $x = \frac{1}{3}$ _____

Write an equation in point-slope form for each line according to the given information.

16. containing $(2, 3)$ and perpendicular to $y = 2x - 1$ _____

17. containing $(1, -3)$ and perpendicular to $y = -3$ _____

18. containing $(3, 4)$ and perpendicular to $2x - 3y = -6$ _____

19. containing $(4, 1)$ and perpendicular to $\frac{1}{2}x + y = 3$ _____

20. containing $(-1, 6)$ and perpendicular to $x = -1$ _____

Reteaching Masters — Algebra 1

NAME _____ CLASS _____ DATE _____

Reteaching
6.1 Solving Inequalities

◆ **Skill A** Solving inequalities that contain addition and subtraction

Recall **Subtraction Property of Inequality:** If equal amounts are subtracted from the expressions on each side of an inequality, the resulting inequality is still true.

Addition Property of Inequality: If equal amounts are added to the expressions on each side of an inequality, the resulting inequality is still true.

◆ **Example 1**
Solve the inequality $x + 4 < 14$.

◆ **Solution**
$x + 4 < 14$
$x + 4 - 4 < 14 - 4$ Subtract 4 from each side of the inequality.
$x < 10$ Simplify.

◆ **Example 2**
Solve the inequality $x - 7 > -13$.

◆ **Solution**
$x - 7 > -13$
$x - 7 + 7 > -13 + 7$ Add 7 to each side of the inequality.
$x > -6$

Solve each inequality.

1. $x - 17 > 43$ _____

2. $11 \leq m - 14.5$ _____

3. $-8 + m \geq -7$ _____

4. $a + 4 < -4$ _____

5. $z + 1 \leq 5$ _____

6. $-3 \geq b + 11$ _____

7. $x + (-8) > 22$ _____

8. $10 \geq -3 + b$ _____

9. $y - 18 > -3$ _____

10. $255.6 + s > 322.7$ _____

11. $z + 9\frac{1}{2} \leq 3\frac{2}{3}$ _____

12. $x - 44,500 > 16,950$ _____

Algebra 1   Reteaching Masters   **63**

NAME _____ CLASS _____ DATE _____

> ◆ **Skill B** Writing inequalities to represent a given situation
>
> **Recall** **Statements of inequality**
> | | |
> |---|---|
> | a is less than b. | $a < b$ |
> | a is greater than b. | $a > b$ |
> | a is less than or equal to b. | $a \leq b$ |
> | a is greater than or equal to b. | $a \geq b$ |
> | a is not equal to b. | $a \neq b$ |
>
> ◆ **Example 1**
> Write an inequality to represent the temperature, T, on a day when the high temperature for that day is 95°F.
>
> ◆ **Solution**
> $T \leq 95$; the temperature was less than or equal to 95°F throughout the day.
>
> ◆ **Example 2**
> Write an inequality to represent the humidity, H, on a day when the humidity reached a high of 68% by midafternoon and dropped to a low of 34% by nightfall.
>
> ◆ **Solution**
> $0.34 \leq H \leq 0.68$; the humidity was between 34% and 68% throughout the day.

13. The maximum possible bowling score for two games is 600 points. If Ed never scores below 100 points for any game, write an inequality to describe his total score, S, for two games.

14. A classroom can seat a maximum of 30 students in 6 rows, with 5 desks in each row. On the first day of school, all of the front desks are occupied, but the classroom is not full. Write an inequality that models this situation. Use S as the number of students present on the first day of school.

15. The scores on a test range from 45 points to 99 points. Write an inequality that represents what one of the scores, S, might be.

16. Sylvia is a salesperson at a shoe store. This week Sylvia sold 9 pairs of shoes on Monday, which was her slowest day of the week. She sold the most shoes on Friday when she sold 23 pairs. Write an inequality that uses S to represent the number of pairs of shoes that Sylvia sold on Wednesday.

Reteaching Masters Algebra 1

Reteaching
6.2 Multistep Inequalities

◆ **Skill A** Solving one-step inequalities

Recall When multiplying or dividing each side of an inequality by the same negative number, reverse the sign of the inequality.

◆ **Example**
Solve each inequality.
a. $x - 12 > 8$
b. $\frac{x}{-2} \leq 5$
c. $-3x \geq 21$

◆ **Solution**
a. $x - 12 > 8$
$x - 12 + 12 > 8 + 12$
$x > 20$

b. $\frac{x}{-2} \leq 5$
$(-2)\frac{x}{-2} \geq (-2)5$
$x \geq -10$

c. $-3x \geq 21$
$\frac{-3x}{3} \leq \frac{21}{3}$
$x \leq -7$

Solve the following inequalities:

1. $y - 18 \leq -3$ _____
2. $a + 2 > 10$ _____
3. $-8 + m < -4$ _____
4. $-6b \geq 18$ _____
5. $\frac{r}{3} > -4$ _____
6. $\frac{d}{-8} \leq 5$ _____
7. $x - 5 < 3$ _____
8. $-p \leq 5$ _____
9. $-\frac{1}{5}y \geq -\frac{3}{5}$ _____

Write and solve an inequality for each problem.

10. A number increased by 5 is at least 8. What is the number? _____

11. Claire earns $0.12 per paper to deliver the daily newspaper. How many papers must she deliver in order to earn at least $6? _____

12. Bob can earn no more than $85. How many T-shirts can he sell at $9 per shirt? _____

13. Katarina has $20 to spend. How many greeting cards can she buy if the cards cost $1.65? _____

14. Karl earns $4.75 per hour by working at the movie theater. How many full hours does Karl have to work in order to earn $120? _____

Algebra 1 — Reteaching Masters

NAME _____ CLASS _____ DATE _____

◆ **Skill B** Solving multistep inequalities

Recall To solve a multistep inequality, first add or subtract, and then multiply or divide.

◆ **Example**
Solve each inequality.
a. $10 - 3x > 28$
b. $8x + 12 < 3x - 8$

◆ **Solution**

a. $10 - 3x > 28$
$-3x > 18$
$x < -6$

b. $8x + 12 < 3x - 8$
$5x + 12 < -8$
$5x < -20$
$x < -4$

Solve the following inequalities:

15. $3y - 1 \leq 14$ _____

16. $9a + 5 > 77$ _____

17. $12 - 5m \geq -13$ _____

18. $\frac{b}{4} - 7 < 8$ _____

19. $\frac{x}{-2} + 3 \leq -2$ _____

20. $10p - 11 < 8p + 3$ _____

21. $5 + 6z \geq 3z - 10$ _____

22. $2d + 7 > 8d - 5$ _____

23. $7m - 9 \leq 3m + 7$ _____

24. $2x - 12 > 6x - 20$ _____

Write and solve an inequality for each problem.

25. Eight more than 3 times a number is less than –7. What is the range of numbers?

_____

26. Sam plans to spend no more than $20. Juice costs $2.50 a gallon. If Sam buys snacks for $8, how many gallons of juice can he buy?

_____

27. Sue plans to spend no more than $50. She buys jeans that cost $30. If T-shirts cost $9, how many T-shirts can she buy?

_____

28. The Spanish Club wants to have at least 25 bottles of soda for its party. Rafael and Sally have volunteered to bring an equal number of bottles. If the club already has 7 bottles, how many bottles do Rafael and Sally need to bring?

_____

66 Reteaching Masters   Algebra 1

NAME _____ CLASS _____ DATE _____

Reteaching
6.3 Compound Inequalities

◆ **Skill A** Graphing a solution to a compound inequality.

Recall There are two basic forms of compound inequalities. The form $a < x < b$ represents all points between a and b on the number line. The form $x < a$ or $x > b$ represents all points to the left of a or to the right of b.

◆ **Example**
Graph each compound inequality on a number line.
  a. $-2 < f < 5$ b. $x \leq 0$ or $x > 4$

◆ **Solution**
  a. The solution is all real numbers greater than -2 and less than 5. That is, the solution is all points on the number line between -2 and 5. Points at -2 and 5 are not part of the solution set because the inequality symbol does not include equality.

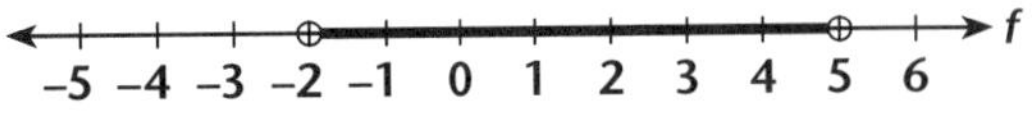

  b. The solution is all real numbers that are less than or equal to 0 or all real numbers that are greater than 4. That is, the solution set is all points at or to the left of 0 or all points to the right of 4. Use a closed circle at 0, but use an open circle at 4.

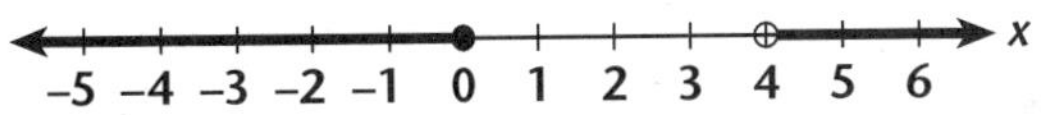

Graph each solution.

1. $x < -4$ or $x \geq 0$

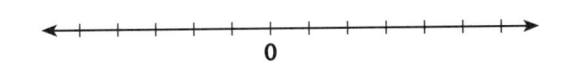

2. $x \leq -4$ or $x > 0$

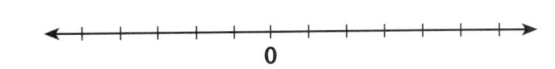

3. $-3\frac{1}{2} < s < 3\frac{1}{2}$

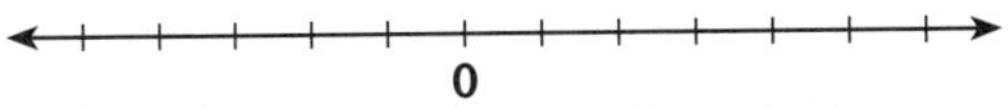

4. $\frac{1}{2} \leq w \leq 4\frac{1}{2}$

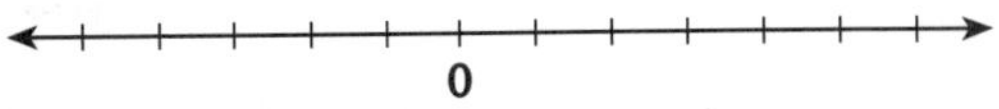

5. $p < -4$ or $p > 4$

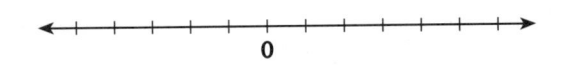

6. $q \leq 3$ or $q \geq 4$

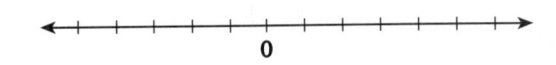

7. $-3.5 \leq d \leq 4\frac{1}{2}$

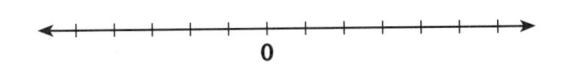

8. $0 < g \leq 3.0$

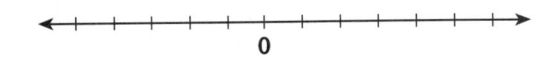

Algebra 1 Reteaching Masters 67

◆ **Skill B** Solving and graphing a solution to a compound inequality

Recall When you solve a compound inequality, you solve two inequalities and then write the solution to the compound inequality based on the solutions to the individual ones.

◆ **Example 1**
Solve $-3.5 \leq 2t + 2.5 < 6.5$ and graph the solution on a number line.

◆ **Solution**

$-3.5 \leq 2t + 2.5 < 6.5$

$-3.5 - 2.5 \leq 2t < 6.5 - 2.5$ Subtract 2.5 from each expression in the inequality.

$-6 \leq 2t < 4$

$3 \leq t < 2$ Divide each expression by 2. You are dividing by a positive number, so keep the inequality symbols as they are.

The solution is all real numbers between -3 and 2, including -3 but not including 2.

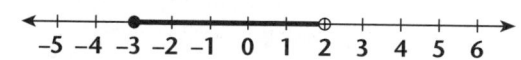

◆ **Example 2**
Solve $-5x \leq -15$ or $2x + 3 < 4$ and graph the solution on a number line.

◆ **Solution**

$-5x \leq -15$

$x \geq 3$ Divide each expression by -5. Reverse the inequality symbol.

or

$2x + 3 < 4$ Subtract 3 from each expression in the inequality.

$2x < 1$

$x < \dfrac{1}{2}$ Divide each side by 2. Keep the inequality symbol as it is.

The solution is all real numbers less than $\dfrac{1}{2}$ or greater than or equal to 3.

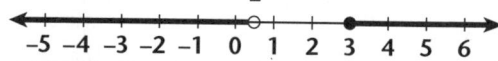

Solve and graph the following compound inequalities:

9. $3.5 \leq 2x + 1.5 < 4.5$ _____

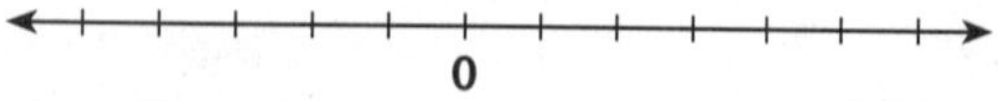

10. $5\dfrac{3}{5} < 3x + 2\dfrac{3}{5} < 8\dfrac{3}{5}$ _____

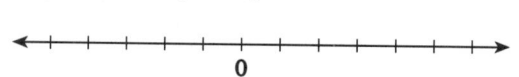

11. $4x + 3.5 < 3\dfrac{1}{2}$ or $2x \geq 7$ _____

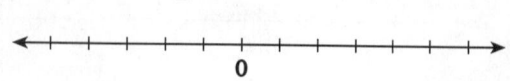

12. $-2x + 1 \geq 5$ or $-3x \leq 9$ _____

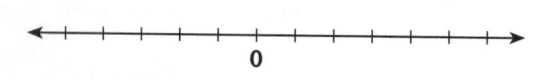

68   **Reteaching Masters**   **Algebra 1**

Reteaching
6.4 Absolute-Value Functions

◆ **Skill A** Determining the absolute value of an expression

Recall The absolute value of a given number is the same number when the number is positive or 0. It is the opposite of the given number when the number is negative.

◆ **Example 1**
  a. Find $|18|$.
  b. Find $|-25|$.

◆ **Solution**
  a. $|18| = 18$
  b. The opposite of -25 is 25. Thus, $|-25| = 25$.

◆ **Example 2**
Find $|4 - 10|$.

◆ **Solution**
$|4 - 10| = |-6| = 6$

Evaluate.

1. $|8|$ _____
2. $|-15|$ _____
3. $|12 - 5|$ _____
4. $|5 - 12|$ _____
5. $|-3 - 6|$ _____
6. $|-6 - 3|$ _____
7. $|8 - (-8)|$ _____
8. $|-8 - 8|$ _____
9. $|8 - 8|$ _____
10. $|35 - 16|$ _____
11. $|16 - 35|$ _____
12. $|-(-25) - 15|$ _____
13. $|15 - (-25)|$ _____
14. $|15 - 25|$ _____

Algebra 1   Reteaching Masters   **69**

NAME _____ CLASS _____ DATE _____

◆ **Skill B** Describing the domain and range of an absolute-value function

Recall The domain of a function is the set of numbers that can be used for the independent variable. The range of a function is the set of numbers that can be used for the dependent variable. When looking at functions, the independent variable is usually x, and the dependent variable is usually y.

◆ **Example 1**
Find the domain and range of $y = 2|x|$.

◆ **Solution**
It is possible to find the absolute value of any real number, so the domain is all real numbers. The absolute value of x is always positive or 0. In this function, every number is multiplied by 2, so this will still result in positive numbers and 0 or all non-negative real numbers.

◆ **Example 2**
Find the domain and range of $y = |x| - 1$.

◆ **Solution**
It is possible to find the absolute value of any real number, so the domain is all real numbers. The range includes -1 because when $x = 0$, $y = -1$. The expression $|x|$ can never be any less than 0 so the range is all numbers greater than or equal to -1, or $y \geq -1$.

Find the domain and range of each function.

15. $y = 5|x|$

16. $y = |x - 6|$

17. $y = -3|x|$

18. $y = |x| - 3$

19. $y = -|x| + 1$

20. $y = |x + 5|$

21. $y = |x| + 6$

22. $y = |x - 6|$

NAME _____ CLASS _____ DATE _____

Reteaching
6.5 Absolute-Value Equations and Inequalities

◆ **Skill A** Solving absolute-value equations

Recall To solve absolute-value equations, you must consider two cases.
Case 1: Consider the quantity within the absolute-value sign as positive.
Case 2: Consider the quantity within the absolute-value sign as negative.

◆ **Example 1**
Solve $|x + 1| = 4$.

◆ **Solution**

Case 1: when $(x + 1)$ is positive
$$x + 1 = 4$$
$$x = 3$$

Case 2: when $(x + 1)$ is negative
$$-(x + 1) = 4$$
$$-x - 1 = 4$$
$$-x = 5$$
$$x = -5$$

Thus, $x = 3$ or $x = -5$.

◆ **Example 2**
Solve $|5x + 7| = 42$.

◆ **Solution**

Case 1: when $(5x + 7)$ is positive
$$5x + 7 = 42$$
$$5x = 35$$
$$x = 7$$

Case 2: when $(5x + 7)$ is negative
$$-(5x + 7) = 42$$
$$-5x - 7 = 42$$
$$-5x = 49$$
$$x = -9.8$$

Thus, $x = 7$ or $x = -9.8$.

Solve each equation if possible. Check your answers.

1. $|x + 2| = 10$ _____

2. $|x - 9| = 5$ _____

3. $|3 - x| = 2$ _____

4. $|x - 12| = 3$ _____

5. $|5 - x| = 1$ _____

6. $|x + 7| = 18$ _____

7. $|2x - 1| = 11$ _____

8. $|8 - 3x| = 1$ _____

9. $|6x + 3| = 27$ _____

10. $|\frac{1}{2}x + 4| = 5$ _____

11. $|5x - 8| = 12$ _____

12. $|-1 - 4x| = 11$ _____

Algebra 1   Reteaching Masters   **71**

NAME _____ CLASS _____ DATE _____

◆ **Skill B** Solving absolute-value inequalities

Recall When you multiply or divide each side of the inequality by a negative number reverse the inequality symbol.

◆ **Example**
Solve $|4 - x| \geq 6$ and graph the solution on a number line.

◆ **Solution**

Case 1: $4 - x \geq 6$ Case 2: $-(4 - x) \geq 6$
$\quad\quad\quad -x \geq 2$ $-4 + x \geq 6$
$\quad\quad\quad\;\; x \leq -2$ $x \geq 10$

Thus, $x \leq -2$ or $x \geq 10$.

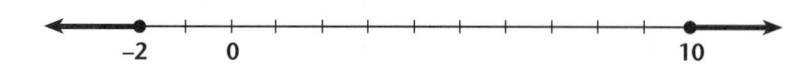

Use shaded circles to show that -2 and 10 are part of the solution.

Solve each inequality and graph each solution on the number line provided.

13. $|x - 1| > 8$

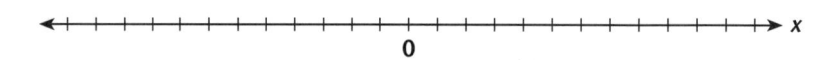

14. $|x + 3| \geq 5$

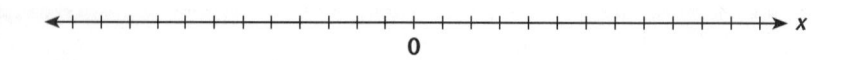

15. $|x - 6| < 2$

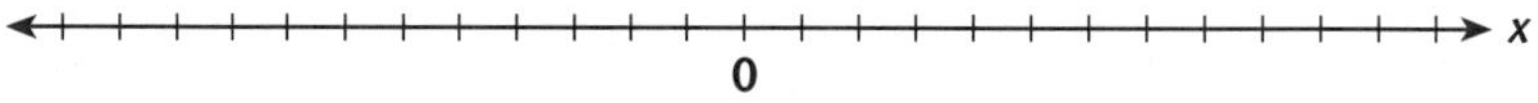

16. $|x + 5| \leq 1$

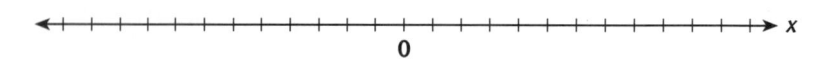

17. $|2 - x| < 3$

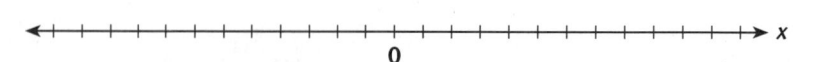

18. $|7 - x| > 3$

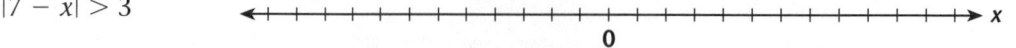

72   **Reteaching Masters**   **Algebra 1**

NAME _____ CLASS _____ DATE _____

Reteaching
7.1 Graphing Systems of Equations

◆ **Skill A** Solving systems of equations by graphing

Recall Write each equation in slope-intercept form, $y = mx + b$. Then graph both equations on the same coordinate plane. The solution is the point where the lines intersect.

◆ **Example**
Solve the system of equations.
$$\begin{cases} y = -x + 2 \\ 2x - y = 1 \end{cases}$$

◆ **Solution**
The first equation is already solved for y. Solve the second equation for y.

$2x - y = 1$
$2x - (2x) - y = 1 - (2x)$ Subtraction Property of Equality
$\dfrac{-y}{-1} = \dfrac{(1 - 2x)}{-1}$ Division Property of Equality
$y = -1 + 2x$ Simplify.
$y = 2x - 1$ Commutative Property of Addition

The graphs of both equations are shown.

The lines intersect at (1, 1), so the solution is (1, 1).

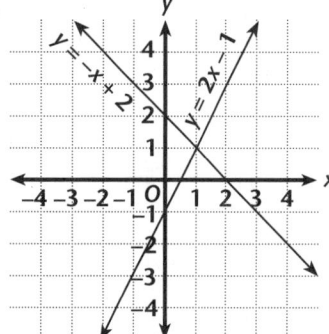

Solve by graphing. Check by substituting your solution into the original equations.

1. $\begin{cases} y = x \\ x + y = 4 \end{cases}$

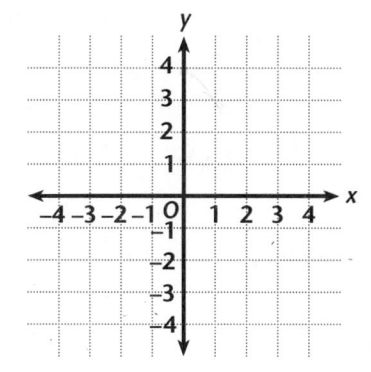

2. $\begin{cases} x + y = 3 \\ y = 2x \end{cases}$

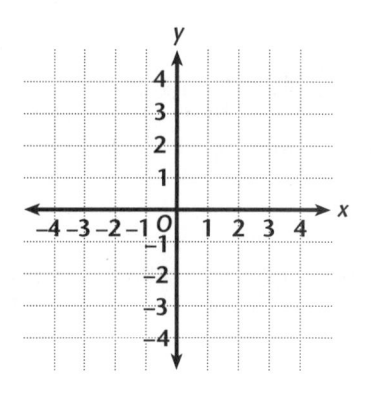

Algebra 1   Reteaching Masters   **73**

NAME _____ CLASS _____ DATE _____

◆ Skill B Finding the approximate solution to a system of equations by graphing

Recall Many real-world problems do not have whole-number solutions. These types of problems can be solved by finding an approximate solution. In order to find an approximate solution, graph the system of equations on one graph and estimate the point of intersection. This point of intersection is the approximate solution.

◆ Example
A cab ride costs $2.50 plus $0.75 per mile traveled if you use Speedy Cab or $1.25 plus $1.70 per mile traveled if you use Comfort Cab. At what distance will the cab rides cost the same amount?

◆ Solution
Let y represent the cost of the cab ride, and let x represent the distance in miles.
Set up a system of equations.   Speedy Cab: $y = 0.75x + 2.5$
                                Comfort Cab: $y = 1.7x + 1.25$
The graphs of both equations are shown.

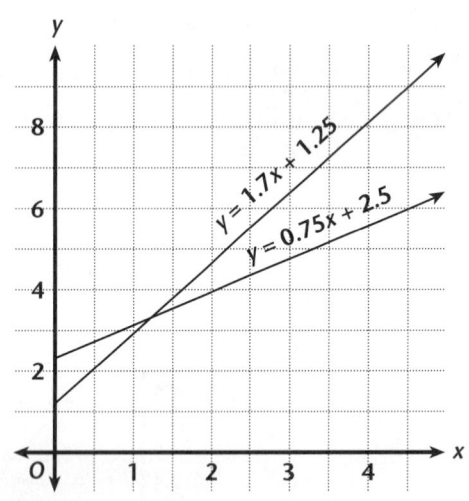

The point of intersection is approximately (1.3, 3.5).
Thus, the solution to the problem is approximately 1.3 miles.

Write a system of equations for each problem. Use your own grid paper to solve each system by graphing.

3. A hot air balloon is 20 feet above the ground and is rising at a rate of 15 feet per minute. Another balloon is 150 feet above the ground and is descending at a rate of 20 feet per minute. At what height will the balloons be the same distance from the ground?

4. The perimeter of a rectangular swimming pool is 40 meters. The length of the pool is 2 meters less than twice its width. What are the dimensions of the pool?

74 **Reteaching Masters**   **Algebra 1**

Reteaching
7.2 The Substitution Method

◆ **Skill A** Solving a system of equations by using substitution

Recall Solve one of the equations for one of the variables. Then substitute the right side of the new equation for that variable in the other equation. The result will be an equation with only one variable. Solve for this variable. Then find the solution for the remaining variable in one of the original equations.

◆ **Example**
Solve by using substitution.
$\begin{cases} x + 6y = 1 \\ 3x - 10y = 31 \end{cases}$

◆ **Solution**
Solve for x in the first equation.
$x + 6y = 1$
$x = 1 - 6y$

Substitute $1 - 6y$ for x in the second equation.
$3x - 10y = 31$
$3(1 - 6y) - 10y = 31$
$3 - 18y - 10y = 31$
$3 - 28y = 31$
$-28y = 28$
$y = -1$

Given $y = -1$, solve for x.
$x + 6y = 1$
$x + -6(-1) = 1$
$x = 7$

Thus, the solution is $(7, -1)$.

Check the solution in both of the original equations in order to make sure that your answer is correct.

Solve by using substitution, and check your answers.

1. $\begin{cases} y = 4x \\ 4x + 5y = -24 \end{cases}$ _____

2. $\begin{cases} y = 2 \\ 2x - 4y = 1 \end{cases}$ _____

3. $\begin{cases} 2x - 4y = 20 \\ \frac{x}{3} - y = 3 \end{cases}$ _____

4. $\begin{cases} 3y = 3x - 3 \\ 3x + 3y = 9 \end{cases}$ _____

5. $\begin{cases} -3x + y = -4 \\ -2x + 3y = 9 \end{cases}$ _____

6. $\begin{cases} -x + 2y = -5 \\ -3x + 5y = -8 \end{cases}$ _____

Algebra 1   Reteaching Masters   75

NAME _____ CLASS _____ DATE _____

◆ **Skill B** Solving word problems using substitution

Recall Define the variables. Then write and solve the resulting system of equations.

◆ **Example**
The sum of two angles is 180°. The measure of the larger angle is 124° more than the measure of the smaller angle. Find the measure of each angle.

◆ **Solution**
Let x represent the measure of the larger angle and let y represent the measure of the smaller angle.

Write a system of equations: $\begin{cases} x + y = 180 \\ y = x + 124 \end{cases}$.

Solve the system by substitution. Substitute the value of y from the second equation for y in the first equation. Then solve for x.

$x + (x + 124) = 180$
$2x + 124 = 180$
$2x = 56$
$x = 28$

Given $x = 28$, solve for y.
$y = x + 124 = 28 + 124 = 152$

The measure of the larger angle is 152°. The measure of the smaller angle is 28°.

Write a system of equations for each problem. Solve by using the substitution method.

7. The sum of two numbers is 346. The smaller number is $\frac{1}{3}$ of the quantity 6 less than the larger number. Find the two numbers.

   _____

   _____

8. The sum of two angles is 90°. The smaller angle is $\frac{2}{3}$ the measure of the larger angle. Find the measures of the two angles.

   _____

   _____

9. Raul is 3 times as old as his sister Sara. The sum of their ages is 28. How old are Raul and Sara? _____

10. If you have $8000 to invest for college and you are investing part at 7% interest and part at 5% interest, how much should you invest at each rate in order to get $500 in interest after one year? _____

Reteaching Masters Algebra 1

Reteaching
7.3 The Elimination Method

♦ **Skill A** Solving a system of equations by using addition or subtraction

Recall If the *x*- or *y*-terms in the two equations are opposites, you can eliminate the variable by using the Addition Property of Equality. If they have the same coefficients, you can eliminate the variable by using the Subtraction Property of Equality.

♦ **Example**
Solve by elimination.
$\begin{cases} x - y = 8 \\ x + y = 2 \end{cases}$

♦ **Solution**
Since y and $-y$ are opposites, use the Addition Property of Equality to combine equations. Then solve the resulting equation for x.

$x - y = 8$
$x + y = 2$
$2x = 10$
$x = 5$

Given $x = 5$, solve for y in either equation.
$5 + y = 2$
$y = -3$

Thus, the solution is $(5, -3)$.

Check the solution in both of the original equations in order to make sure that your answer is correct.

Solve each system of equations by elimination, and check your solution.

1. $\begin{cases} 2x - 2y = -10 \\ 2x + 2y = 50 \end{cases}$

2. $\begin{cases} x + 3y = 5 \\ x + 2y = 3 \end{cases}$

3. $\begin{cases} 5x + y = 9 \\ -5x + y = 7 \end{cases}$

4. $\begin{cases} 3x - 12y = 18 \\ 9x + 12y = 30 \end{cases}$

5. $\begin{cases} 3x + 4y = 2 \\ 4x - 4y = 12 \end{cases}$

6. $\begin{cases} \frac{1}{2}x + y = 5 \\ \frac{3}{2}x - y = 30 \end{cases}$

Algebra 1   Reteaching Masters

NAME _____ CLASS _____ DATE _____

◆ **Skill B** Solving a system of equations by using multiplication

Recall If you multiply each side of an equation by the same number, the products are equal.

◆ **Example**
Solve by elimination. $\begin{cases} x + y = 6 \\ 2x - 3y = 2 \end{cases}$

◆ **Solution**
Multiply each side of the first equation by 2.
$$2(x + y) = 2(6)$$
$$2x + 2y = 12$$

This will result in equal but opposite x-terms in both equations.
$$2x + 2y = 12$$
$$2x - 3y = 2$$

Use the Subtraction Property of Equality to solve for y.
$$2y - (-3y) = 12 - 2$$
$$5y = 10$$
$$y = 2$$

Use $y = 2$ and the first equation to solve for x.
$$x + y = 6$$
$$x + 2 = 6$$
$$x = 4$$

Thus, the solution is (4, 2).

Check the solution in both of the original equations in order to make sure that your answer is correct.

Solve each system of equations by elimination and check your solution.

7. $\begin{cases} 2x - 10y = 0 \\ 4x - 6y = 14 \end{cases}$ _____

8. $\begin{cases} 4x + y = 8 \\ x - 10y = 2 \end{cases}$ _____

9. $\begin{cases} 3x - 8y = 9 \\ x - y = 2 \end{cases}$ _____

10. $\begin{cases} 5x - 4y = 2 \\ 2x + y = 6 \end{cases}$ _____

11. $\begin{cases} 3x - 2y = 10 \\ 2x - y = 8 \end{cases}$ _____

12. $\begin{cases} 4x + 10y = 6 \\ -2x + 6y = -14 \end{cases}$ _____

Write and solve a system of equations for the following problem.

13. The drama club sold T-shirts and baseball caps in a fund-raiser. The T-shirts sold for $12, and the caps sold for $9. The club sold a total of 114 shirts and caps. If the club raised $1242, how many T-shirts and how many caps were sold? _____

78   Reteaching Masters   Algebra 1

NAME _____ CLASS _____ DATE _____

Reteaching
7.4 Consistent and Inconsistent Systems

◆ **Skill A** Solving inconsistent systems

Recall The graphs of the equations in a system of equations can intersect in one point, in an infinite number of points, or in no points. If the system has one solution or infinitely many solutions, the system is called consistent. If there is no solution, the system is called inconsistent.

◆ **Example 1**
Is the system $\begin{cases} y = x - 1 \\ y = -x + 3 \end{cases}$ consistent?

◆ **Solution**
Graph the system.

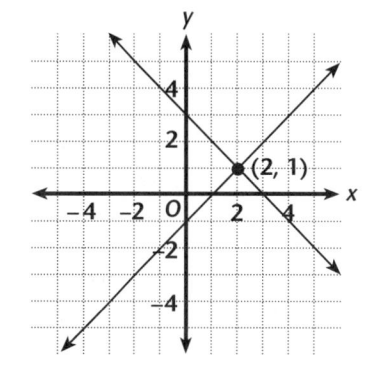

From the graphs, you can see that the lines intersect. Thus, the system is consistent.

If you solve the system by using the Addition Property of Equality, the result will be $x = 2$ and $y = 1$. That is, the solution is (2, 1), so the system is consistent.

◆ **Example 2**
Is the system $\begin{cases} y = 2x + 4 \\ y = 2x - 3 \end{cases}$ consistent?

◆ **Solution**
Graph the system.

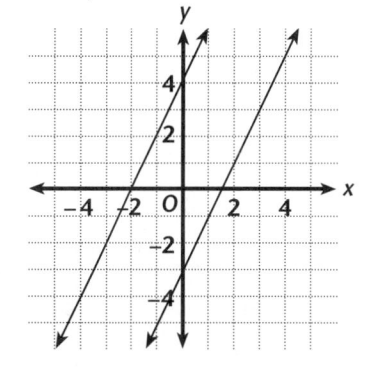

The lines are parallel. If you solve the system by using the Subtraction Property of Equality, the result is $0 = 1$.

There are no solutions to the system. The system is inconsistent.

Solve each system algebraically. Identify each system as consistent or inconsistent.

1. $\begin{cases} 2y - 3x = 2 \\ 2y = 3x - 4 \end{cases}$ _____

2. $\begin{cases} 2x + 3y = 9 \\ 2x - 3y = 3 \end{cases}$ _____

3. $\begin{cases} y = 5x + 2 \\ y = -5x + 2 \end{cases}$ _____

4. $\begin{cases} y = x + 2 \\ x = y + 4 \end{cases}$ _____

Algebra 1   Reteaching Masters

NAME _____ CLASS _____ DATE _____

◆ Skill B Solving dependent systems

Recall There are two types of consistent systems, independent and dependent. An independent system has exactly one solution. A dependent system has infinitely many solutions. If a system is dependent, the graphs of both equations are the same line.

◆ **Example**
Solve the system algebraically.
$\begin{cases} 2x - y = 5 \\ 4x - 2y = 10 \end{cases}$

Identify the system as consistent or inconsistent. If the system is consistent, determine whether it is independent or dependent.

◆ **Solution**
Solve the system by elimination.
Multiply the first equation by 2. $4x - 2y = 10$
Write the second equation as it is. $4x - 2y = 10$

When you apply the Subtraction Property of Equality, the result is $0 = 0$. Any ordered pair that solves the first equation will also solve the second equation.

There are infinitely many solutions to the system. Thus, the system is consistent and dependent.

Solve each system algebraically. Identify each system as consistent and dependent, consistent and independent, or inconsistent.

5. $\begin{cases} 4y = -2x - 6 \\ 2y = -x - 3 \end{cases}$

6. $\begin{cases} x + y = 4 \\ x + y = 2 \end{cases}$

7. $\begin{cases} 5x + 3y = -6 \\ 2x + y = -4 \end{cases}$

8. $\begin{cases} x + 3y = 1 \\ 2x + 6y = 3 \end{cases}$

9. $\begin{cases} 3x + 6y = 3 \\ 2x + 4y = 2 \end{cases}$

10. $\begin{cases} 6x - 9y = 12 \\ 8x - 12y = 16 \end{cases}$

11. $\begin{cases} 2x - y = 2 \\ 3x - 2y = 1 \end{cases}$

12. $\begin{cases} 3x + y = 2.7 \\ x - 2y = -1.9 \end{cases}$

NAME _____ CLASS _____ DATE _____

Reteaching
7.5 Systems of Inequalities

◆ **Skill A** Graphing systems of linear inequalities

Recall A system of linear inequalities is graphed in much the same way as a system of equations. Solve each inequality for y and then graph the inequalities as solid or dotted lines on the same coordinate plane. Shade the region that contains the solutions to both inequalities.

◆ **Example**
Graph. $\begin{cases} 3x + y > 8 \\ x + y \leq 4 \end{cases}$

◆ **Solution**
Solve both inequalities for y.

$3x - 3x + y > 8 - 3x$ Subtraction Property of Equality
$\qquad\qquad y > 8 - 3x$ Simplify.
$x - x + y \leq 4 - x$ Subtraction Property of Equality
$\qquad\qquad y \leq 4 - x$ Simplify.

Graph both inequalities on the same coordinate plane.

The first line, $y > 8 - 3x$, is dotted because the solutions do not include the line. The other line, $y \leq 4 - x$, is solid because it is included in the solution.

The solutions lie in the shaded region between the two lines and below the point of intersection.

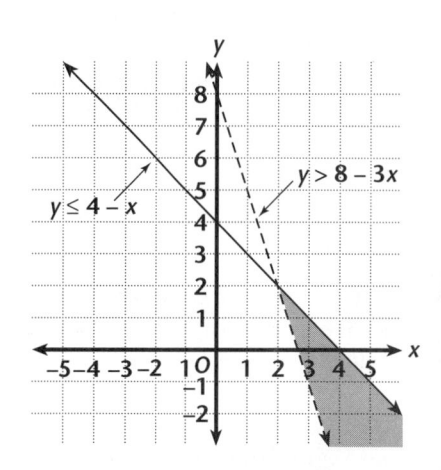

Graph each system of inequalities on the grid provided.

1. $\begin{cases} y < 8x - 4 \\ y > x - 2 \end{cases}$

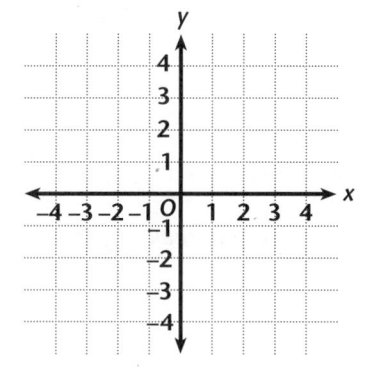

2. $\begin{cases} y - 2x \geq 3 \\ 2y + x \geq -5 \end{cases}$

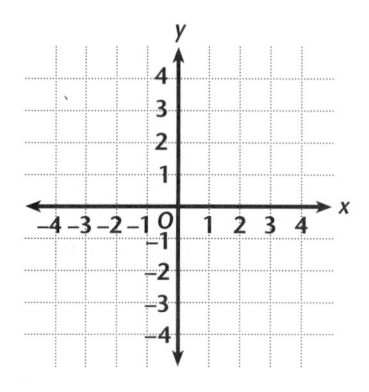

Algebra 1   Reteaching Masters

NAME _____ CLASS _____ DATE _____

◆ **Skill B** Solving systems of linear inequalities by graphing

Recall The solution is the intersection of the graphs of the solutions to each inequality.

◆ **Example**
To ensure a growing season of sufficient length, ABC Agriculture has at most 16 days to plant wheat and corn. They can plant corn at a rate of 10 acres per day and wheat at a rate of 15 acres per day. If they have no more than 200 acres available, how many acres of each type of crop can they plant?

◆ **Solution**
Let c represent the number of days that corn can be planted and let w represent the number of days that wheat can be planted.

Write a system of inequalities.
$c + w \leq 16$
$10c + 15w \leq 200$

Solve both inequalities for c.
$c \leq 16 - w$
$c \leq 20 - \frac{3}{2}w$

Then graph both inequalities. The solution is shown in the shaded area. There are many possible solutions, and one of these is (8, 8). ABC Agriculture could plant corn for 8 days.

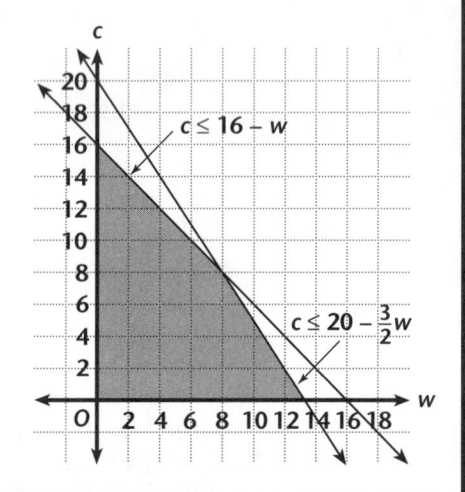

Write a system of inequalities for each problem. Solve each system by graphing on your own grid paper. List at least two possible solutions.

3. You are going camping, and you want to make a trail mix. The most that you want to spend is $30 for candy and raisins. The store sells candy in the bulk section for $4 per pound and raisins for $2 per pound. If you need to have at least 8 pounds of mix, how much of each type can you buy?

_____

_____

4. The senior class needs at least $1000 for prom. They decide to have a car wash in order to raise money, charging $4 for small vehicles and $6 for large vehicles. If they can wash no more than 200 vehicles, how many vehicles of each type must they wash in order to raise at least $1000?

_____

_____

82 Reteaching Masters Algebra 1

Reteaching
7.6 Classic Puzzles in Two Variables

◆ **Skill A** Solving age problems

Recall To solve an age problem, represent a person's age in the future or in the past.

◆ **Example**
Marge is 3 years older than Susan. Four years ago, Marge was twice as old as Susan. How old is Marge now?

◆ **Solution**
Let m represent Marge's age now and let s represent Susan's age now.

Write an equation comparing the ages of the girls now: $m = s + 3$.
Write an equation comparing their ages 4 years ago: $m - 4 = 2(s - 4)$.

Solving the system $\begin{cases} m = s + 3 \\ m - 4 = 2(s - 4) \end{cases}$ gives $s = 7$ and $m = 10$.

Thus, Marge is now 10 years old.

Write a system for each age problem. Then solve each problem.

1. Joel is 1 year younger than Roberto. The sum of their ages is 27. How old is each?

2. Latisha is three times as old as her brother. In six years she will be twice as old as her brother. How old is Latisha now?

◆ **Skill B** Solving coin problems

Recall To solve a coin problem, write one equation relating the coins and another equation relating the values of the coins.

◆ **Example**
Paul has 25 dimes and nickels. He has a total of $2.00.
How many dimes and nickels does Paul have?

◆ **Solution**
Let d represent dimes and let n represent nickels.
Paul has 25 coins. Thus, $d + n = 25$.
The dimes are worth $10d$ cents and the nickels are worth $5n$ cents.
He has $2.00, or 200 cents. Thus, $10d + 5n = 200$.
Solving the system $\begin{cases} d + n = 25 \\ 10d + 5n = 200 \end{cases}$ gives $d = 15$ and $n = 10$.
Thus, Paul has 15 dimes and 10 nickels.

Algebra 1   Reteaching Masters   83

NAME _____ CLASS _____ DATE _____

Write a system for each coin problem. Then solve each problem.

3. Raul has $4.32 in pennies and nickels. If Raul has a total of 260 coins, how many of each does he have? _____

4. Ronnie has 57 nickels and quarters. If Ronnie has $7.25, how many nickels and quarters does he have? _____

5. Rachel bought 42 stamps for $14.40. How many 32-cent stamps and 40-cent stamps did Rachel buy? _____

◆ **Skill C** Solving number-digit problems

Recall To write an equation relating the values of the digits, write the number in expaned form. For example, $32 = 3(10) + 2$.

◆ **Example 1**
The sum of the digits of a two-digit number is 11. The ones digit is less than twice the tens digit. What is the number?

◆ **Solution**
Let t represent the tens digit and let u represent the units digit.
Write a system relating the digits of the number.
Solving $\begin{cases} t + u = 11 \\ u = 2t - 1 \end{cases}$ gives $t = 4$ and $u = 7$.
Thus, the number is 47.

◆ **Example 2**
The sum of the digits of a two-digit number is 14. By reversing the digits, the number is increased by 18. Find the original number.

◆ **Solution**
Let t represent the tens digit and u represent the units digit. Then $t + u = 14$.
$10t + u$ represents a two digit number. $10u + t$ represents the number when the digits are reversed. Thus, $10t + u = (10u + t) + 18$.
Solving the system $\begin{cases} t + u = 14 \\ 9t - 9u = 18 \end{cases}$ gives $t = 6$ and $u = 8$.
Thus, the number is 68.

Write a system for each digit problem. Then solve the problem.

6. The sum of the digits of a two-digit number is 12. The ones digit is 3 times greater than the tens digit. What is the number? _____

7. The sum of the digits of a two-digit number is 10. If you reverse the digits, the new number is one more than twice the original number. Find the original number. _____

84 Reteaching Masters   Algebra 1

Reteaching

8.1 Laws of Exponents: Multiplying Monomials

◆ **Skill A** Understanding exponents and powers

Recall In the expression 2^4, 2 is called the base, and 4 is called an exponent. The exponent tells how many times to multiply the base by itself. Thus, $2^4 = 2 \cdot 2 \cdot 2 \cdot 2 = 16$. 16 is the fourth power of 2.

◆ **Example 1**
Find the value of 3^5.

◆ **Solution**
The base 3 is multiplied by itself 5 times.
$$3^5 = 3 \cdot 3 \cdot 3 \cdot 3 \cdot 3 = 243$$

◆ **Example 2**
Find the value of 5^1.

◆ **Solution**
Any number raised to the power of 1 is the number itself. Thus, $5^1 = 5$.

Recall When the base is 10, the exponent also tells you how many zeros to place after 1 in the answer.

◆ **Example 3**
Find the value of 10^3.

◆ **Solution**
The base 10 is multiplied by itself 3 times. The value of 10^3 is 1 followed by 3 zeros. Thus, $10^3 = 1000$.

Find the value of each expression.

1. 5^4 _____

2. 9^2 _____

3. 10^6 _____

4. 2^7 _____

5. 8^5 _____

6. 3^6 _____

7. 6^1 _____

8. 10^4 _____

9. 12^2 _____

10. 1^8 _____

Algebra 1   Reteaching Masters

NAME _____ CLASS _____ DATE _____

◆ **Skill B**   Using the Product-of-Powers Property to simplify expressions

Recall   If x is any number and m and n are any positive integers, then: $x^m \cdot x^n = x^{m+n}$.

◆ **Example**
Simplify.

◆ **Solution**
$2^3 \cdot 2^4 = 2^{3+4}$
$ = 2^7$
Thus, $2^3 \cdot 2^4 = 2^7$.

Simplify each product. Then find the value of the expression.

11. $3^5 \cdot 3^4$ _____   12. $2^5 \cdot 2^2$ _____

13. $10^5 \cdot 10^3$ _____   14. $5^6 \cdot 5^1$ _____

15. $8^4 \cdot 8^6$ _____   16. $4^3 \cdot 4^4$ _____

◆ **Skill C**   Using the Product-of-Powers Property to multiply monomials

Recall   To multiply two monomials, multiply the constants and multiply the variables with the same base.

◆ **Example**
Simplify $(3m^4n)(-2m^2n)$.

◆ **Solution**
$(3m^4n)(-2m^2n) = (3 \cdot -2)(m^4 \cdot m^2)(n \cdot n)$
$ = -6m^6n^2$

Simplify each product.

17. $(5a^2)(3a^3)$ _____   18. $(-7cd^2)(3c^2)$ _____

19. $(-s^3t)(-5t^4)$ _____   20. $(6p^5)(4p^2q^3)$ _____

21. $(m^3n^2)(4m^2n^2)$ _____   22. $(a^2b^3)(2b^2c^2)(3a^4)$ _____

86   Reteaching Masters   Algebra 1

NAME _____ CLASS _____ DATE _____

Reteaching
8.2 Laws of Exponents: Powers and Products

◆ **Skill A** Raising a power to a power

Recall In the expression $(2^3)^2$, 2^3 is raised to the second power. Because 2^3 is used as a factor two times, $(2^3)^2 = 2^3 \cdot 2^3 = 2^6$. When you simplify a power of a power, the exponents are multiplied.

◆ **Example 1**
Find the value of $(3^3)^2$.

◆ **Solution**
$(3^3)^3 = 3^{3 \cdot 3}$
$= 3^9$
$= 19{,}683$

Recall If x is any number and m and n are any positive integers, then $(x^m)^n = x^{mn}$.

◆ **Example 2**
Simplify each expression.
a. $(b^2)^5$ b. $(2q^2)^4$

◆ **Solution**
a. $(b^2)^5 = b^{2 \cdot 5}$
$= b^{10}$

b. $(2q^2)^4 = (2^{1 \cdot 4})(q^{2 \cdot 4})$
$= 2^4 q^8$
$= 16 q^8$

Find the value of each numerical expression. Simplify each algebraic expression.

1. $(2^3)^4$ _____   2. $(3^2)^3$ _____

3. $(y^4)^3$ _____   4. $(m^5)^2$ _____

5. $(2s^3)^2$ _____   6. $5(r^5)^2$ _____

7. $(3c^6)^3$ _____   8. $(n^3)^d$ _____

Algebra 1 — Reteaching Masters 87

NAME _____ CLASS _____ DATE _____

◆ **Skill B** Raising a monomial to a power

Recall When a monomial is raised to a power, raise each term to that power. $(ab)^2$ means $(ab)(ab)$. Thus, $(ab)^2 = ab \cdot ab = (a \cdot a)(b \cdot b) = a^2b^2$.

If x and y are any numbers and n is a positive integer, then $(xy)^n = x^n y^n$.

◆ **Example**
Simplify $(3b^2c^3)^2$.

◆ **Solution**
$(3b^2c^3)^2 = (3^2)(b^2)^2(c^3)^2 = 9b^4c^6$

Simplify each expression.

9. $(10^3)^2$ _____

10. $(2y^3)^5$ _____

11. $(6x^4)^3$ _____

12. $(8q^3)^3$ _____

13. $(c^2d^2)^4$ _____

14. $(9mn^5)^2$ _____

15. $4(e^4f)^3$ _____

16. $(2p^5r^3)^4$ _____

◆ **Skill C** Finding powers of -1

Recall All even powers of -1 are equal to 1.
All odd powers of -1 are equal to -1.

◆ **Example**
Simplify each of the following:
   a. $(-c)^3$ **b.** $(-2p)^4$

◆ **Solution**
   a. $(-c)^3 = (-1 \cdot c)^3 = (-1)^3(c)^3 = -1(c^3) = -c^3$
   b. $(-2p)^4 = (-2)^4(p)^4 = 16p^4$

Simplify each expression.

17. $(-3y^3)^2$ _____

18. $(-gh^4)^5$ _____

19. $(ab^2)^3(-ab^3)$ _____

20. $(-2c^2d^3)^4(-3cd^2)^3$ _____

88   **Reteaching Masters**   **Algebra 1**

NAME _____ CLASS _____ DATE _____

Reteaching
8.3 Laws of Exponents: Dividing Monomials

◆ **Skill A** Using the Quotients-of-Powers Property

Recall In a quotient of powers, one power is divided by another. $\frac{2^5}{2^3}$ means $\frac{2 \cdot 2 \cdot 2 \cdot 2 \cdot 2}{2 \cdot 2 \cdot 2}$, so $\frac{2^5}{2^3} = 2 \cdot 2 = 4$.

◆ **Example 1**
Find the value of $\frac{3^7}{3^3}$.

◆ **Solution**
$\frac{3^7}{3^3} = \frac{3 \cdot 3 \cdot 3 \cdot 3 \cdot 3 \cdot 3 \cdot 3}{3 \cdot 3 \cdot 3} = 3 \cdot 3 \cdot 3 \cdot 3 = 81$

Recall In the example above, the power of the quotient is the difference between the powers of the numerator and the denominator.

If x is any number except 0 and m and n are any positive integers, where $m > n$, then $\frac{x^m}{x^n} = x^{m-n}$.

◆ **Example 2**
Simplify the expression $\frac{20z^8}{4z^3}$.

◆ **Solution**
$\frac{20z^8}{4z^3} = 5(z^{8-3}) = 5z^5$

Find the value of each numerical expression. Simplify each algebraic expression.

1. $\frac{7^5}{7^2}$ _____

2. $\frac{2^9}{2^4}$ _____

3. $\frac{8d^8}{2d^2}$ _____

4. $\frac{5m^4}{10m}$ _____

5. $\frac{10^{15}}{10^6}$ _____

6. $\frac{-36r^9}{4r^5}$ _____

7. $\frac{-12c^{10}}{-2c^3}$ _____

8. $\frac{9g^8}{-g}$ _____

Algebra 1   Reteaching Masters **89**

NAME _____ CLASS _____ DATE _____

◆ **Skill B** Finding the quotient of monomials

   Recall To divide monomials, divide the constants and the variables with the same base.

   ◆ **Example**
   Simplify $\dfrac{12a^6b^4}{-3a^4b^3}$.

   ◆ **Solution**
   $\dfrac{12a^6b^4}{-3a^4b^3} = \left(\dfrac{12}{-3}\right)(a^{6-4})(b^{4-3}) = -4a^2b$

Simplify each expression.

9. $\dfrac{x^2y^5}{xy^3}$ _____

10. $\dfrac{p^7q^5r^2}{p^3q^4}$ _____

11. $\dfrac{-30g^9h^8}{5g^3h^6}$ _____

12. $\dfrac{24y^8z^5}{-32y^6z}$ _____

13. $\dfrac{-9s^{12}t^9}{-3s^8t^6}$ _____

14. $\dfrac{8.4(b^2c^3)^3}{2b^3c^4}$ _____

◆ **Skill C** Finding the power of a fraction

   Recall If n is a positive number and a and b are numbers, where $b \neq 0$, then $\left(\dfrac{a}{b}\right)^n = \dfrac{a^n}{b^n}$.

   ◆ **Example**
   Simplify $\left(\dfrac{25d^4e^6}{5d^2f}\right)^3$.

   ◆ **Solution**
   $\left(\dfrac{25d^4e^6}{5d^2f}\right)^3 = \left(\dfrac{5d^2e^6}{f}\right) = \dfrac{125d^6e^{18}}{f^3}$

Simplify each expression.

15. $\left(\dfrac{2r^3}{n}\right)$ _____

16. $\left(\dfrac{-35c^2m}{5}\right)^3$ _____

17. $\left(\dfrac{-20c^3}{(-2c)^2}\right)$ _____

18. $\left(\dfrac{d^9e^{12}}{(d^2e^3)^2}\right)^2$ _____

90 Reteaching Masters   Algebra 1

NAME _____ CLASS _____ DATE _____

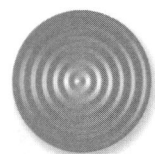

Reteaching
8.4 Negative and Zero Exponents

◆ **Skill A**   Understanding negative and zero exponents

Recall   There is a pattern between exponents and powers of the same base.
As the exponents decrease by 1, 2^4 2^3 2^2 2^1
the value of the power decreases
by a factor of $\frac{1}{2}$.   16   8   4   2

◆ **Example 1**
What is the value of 2^0?

◆ **Solution**
Add 2^0 to the table above and complete the pattern.

2^4 2^3 2^2 2^1 2^0

16   8   4   2   ?

The value of the power decreases by a factor of $\frac{1}{2}$, so $2^0 = 1$.

This pattern will hold true for any base that is a nonzero number, x, or $x^0 = 1$.

◆ **Example 2**
What is the value of 2^{-1}?

◆ **Solution**
Add 2^{-1} to the table above and complete the pattern.

2^4 2^3 2^2 2^1 2^0 2^{-1}

16   8   4   2   1   ?

The value of the power decreases by a factor of $\frac{1}{2}$, so $2^{-1} = \frac{1}{2}$.

If x is any number except zero and n is any integer, then $x^{-n} = \frac{1}{x^n}$.

Evaluate each expression.

1. 4^0 _____   2. 5^{-2} _____

3. 8^0 _____   4. 4^{-1} _____

5. 3^{-3} _____   6. 1^{-2} _____

7. 5^{-3} _____   8. 4^{-3} _____

Algebra 1   Reteaching Masters   **91**

NAME _____ CLASS _____ DATE _____

◆ **Skill B** Simplifying expressions containing negative and zero exponents

Recall To add two integers with the same sign, add their absolute values and keep the common sign. To add two integers with different signs, subtract their absolute values and use the sign of the number with the greater absolute value. Subtraction is the same as adding the opposite.

◆ **Example 1**
Simplify the expression $y^{-5} \cdot y^3$.

◆ **Solution**
To multiply powers of the same base, add the exponents.
$y^{-5} \cdot y^3 = y^{-5+3} = y^{-2}$

◆ **Example 2**
Simplify the expression $\dfrac{m^4}{m^7}$.

◆ **Solution**
To divide powers of the same base, subtract the exponents.
$\dfrac{m^4}{m^7} = m^{4-7} = m^{-3}$

◆ **Example 3**
Simplify the expression $c^{-3} \cdot c^0$.

◆ **Solution**
This expression represents a product of powers of the same base, so the product is found by adding the exponents. Thus, $c^{-3} \cdot c^0 = c^{-3+0} = c^{-3}$. Alternatively, $c^0 = 1$ because any base to the zero power equals 1. A factor multiplied by 1 is itself. Thus, $c^{-3} \cdot c^0 = c^{-3} \cdot 1 = c^{-3}$.

Simplify each expression.

9. $a^3 \cdot a^{-5}$ _____

10. $c^2 \cdot c^{-7}$ _____

11. $\dfrac{y^3}{y^6}$ _____

12. $\dfrac{m^{-3}}{m^6}$ _____

13. $p^8 \cdot p^0$ _____

14. $q^0 \cdot q^{-5}$ _____

15. $x^{-8} \cdot x^{-3}$ _____

16. $z^{-5} \cdot z^8$ _____

17. $\dfrac{t^{-5}}{t^{-10}}$ _____

18. $5^{-3} \cdot 5^8$ _____

19. $x^5 \cdot x^{-3} \cdot x^{-7}$ _____

20. $3^3 \cdot 3^{-10} \cdot 3^6$ _____

21. $\dfrac{t^{-5} \cdot t^5}{t^3}$ _____

22. $\dfrac{4^7}{4^{-3}}$ _____

23. $5^3 \cdot 5^0 \cdot 5^{-1}$ _____

24. $a^2 \cdot a^{-5}$ _____

25. $\dfrac{r^{10} \cdot r^{-2}}{r^5}$ _____

26. $\dfrac{2^{10} \cdot 2^{-10}}{2^{10}}$ _____

92 Reteaching Masters Algebra 1

Reteaching
8.5 Scientific Notation

♦ Skill A Writing numbers in scientific notation

Recall A number is in scientific notation if it is written as a product of two factors. The first factor is a number from 1 to 10, including 1 but not including 10. The second number is a power of 10.

♦ Example 1
Write 2,300,000,000 in scientific notation.

♦ Solution
First place the decimal between the 2 and the 3. The result is the decimal 2.3. Count the number of places that the decimal point was moved in order to form 2.3. Since the decimal point was moved 9 places to the *left*, the exponent of 10 is 9.

$$2,300,000,000 = 2.3 \times 10^9$$

♦ Example 2
Write 0.0000052 in scientific notation.

♦ Solution
First place the decimal between the 5 and the 2. The result is the decimal 5.2. Count the number of places that the decimal was moved in order to form 5.2. Since the decimal point was moved 6 places to the *right*, the exponent of 10 is −6.

$$0.0000052 = 5.2 \times 10^{-6}$$

Write each number in scientific notation.

1. 4,570,000,000 _____
2. 0.0000023 _____

3. 0.00458 _____
4. 62,000,000 _____

5. 70,500,000,000 _____
6. 0.0000875 _____

7. 5800 _____
8. 0.026 _____

9. 35,000,000 _____
10. 0.000000072 _____

11. 2,070,000,000,000 _____
12. 0.00305 _____

Algebra 1 — Reteaching Masters

NAME _____ CLASS _____ DATE _____

◆ **Skill B** Performing computations with scientific notation

Recall A power of 10 written in exponential form indicates the number of zeros in the number when it is written in customary form. That is, 10^4 means 4 zeros in the number 10,000.

◆ **Example 1**
Perform the following computations:
  a. $(7 \times 10^3)(5 \times 10^4)$
  b. $\dfrac{7 \times 10^7}{2 \times 10^3}$

◆ **Solution**
  a. To multiply numbers in scientific notation, first multiply the whole or decimal factors. In this example, $7 \times 5 = 35$. Then multiply the powers of 10 by adding the exponents. $10^3 \cdot 10^4 = 10^7$. The product 35×10^7 should be changed to scientific notation because 35 is not between 1 and 10.
  $$35 \times 10^7 = 3.5 \times 10^1 \times 10^7 = 3.5 \times 10^8$$

  b. To divide numbers in scientific notation, first divide the whole or decimal factors. In this example, $7 \div 2 = 3.5$. Then divide the powers of 10 by subtracting the exponents. $10^7 \div 10^3 = 10^4$. The quotient is 3.5×10^4. This result is in scientific notation because the decimal factor is between 1 and 10.

◆ **Example 2**
Perform the following computations by using a calculator:
  a. $(2.34 \text{ E } 02)(1.7 \text{ E } 03)$
  b. $\dfrac{5.6 \text{ E } 05}{2.8 \text{ E } 02}$

◆ **Solution**
  a. When numbers are written in scientific notation on a calculator, the power of 10 is expressed in E notation, where E is followed by the exponent of 10. 2.34 E 02 means 2.34×10^2. This number can be entered on the calculator by the following succession of keys: 2.34 $\boxed{\text{EE}}$ 2. Enter the multiplication sign and the second number in the same form. The result is 3.978 E 05, or 3.978×10^5.

  b. Enter the numerator in E notation. Use the $\boxed{\text{EE}}$ key for E. Enter the division sign and follow it with the denominator. The quotient is 2 E 03, or 2×10^3.

Perform the following computations. Write your answers in scientific notation.

13. $(8 \times 10^4)(3 \times 10^3)$ _____

14. $\dfrac{6.5 \times 10^5}{1.3 \times 10^3}$ _____

15. $(4.3 \times 10^2)(5.2 \times 10^4)$ _____

16. $\dfrac{8.4 \times 10^6}{2.1 \times 10^2}$ _____

Perform each computation by using a calculator.

17. $(5.63 \text{ E } 03)(4.2 \text{ E } 04)$ _____

18. $\dfrac{7.5 \text{ E } 04}{2.5 \text{ E } 02}$ _____

19. $(4.25 \text{ E } 05)(6.32 \text{ E } 06)$ _____

20. $\dfrac{1.2 \text{ E } 05}{6 \text{ E } 02}$ _____

Reteaching Masters — Algebra 1

NAME _____ CLASS _____ DATE _____

Reteaching
8.6 Exponential Functions

◆ **Skill A**   Finding population growth

Recall   Exponential functions apply to situations in which values in the range of the function change by a fixed rate. One of these situations is population growth.

◆ **Example**
The population of a city grows at a yearly rate of 1.5%. The city has a current population of 500,000. Predict the population of the city two years from now.

◆ **Solution**
After one year, the growth in population is (500,000)(1.5%), or 7500 people. The population is now 500,000 + 7500, or 507,500. After two years, the growth in population is (507,500)(1.5%) or 7612.5. The population two years from now will be 507,000 + 7612.5, or about 514,613.

A city has a current population of 700,000 people and a yearly growth rate of 6.5%. Predict the population after each number of years.

1. 2 years _____   2. 4 years _____

◆ **Skill B**   Using the general growth formula to find population growth

Recall   The general growth formula is $P = A(1 + r)^t$, where P is the amount after t years at a yearly growth rate r expressed as a decimal, and A is the original amount.

◆ **Example**
Use the general growth formula to predict the population discussed in the previous example.

◆ **Solution**
Use the general growth formula. Replace A by the original amount, 500,000. Replace r by the rate written as the decimal, 0.015. Replace t by the number of years, 2.
$P = 500,000(1 + 0.015)^2 = 500,000(1.015)^2 = 515,112.5 \approx 515,113$
The population after two years is expected to be about 515,113 people.

Given the following data, estimate the population of each city.

3. current population: 1,000,000; growth rate: 4.5%; after 5 years _____

4. current population: 350,000; growth rate: 7%; after 3 years _____

5. current population: 8,200,000; growth rate: 11.4%; after 4 years _____

Algebra 1   Reteaching Masters   **95**

◆ Skill C Understanding exponential functions

Recall When a whole-number base is used with a positive integral exponent, the power is greater than or equal to the base. When a whole-number base is used with a nonpositive integral exponent, the power is less than the base.

◆ Example
For each equation,
- describe the effect on y as x increases from zero;
- describe the y-value at $x = 0$;
- describe the effect on y as x decreases from zero; and
- graph the equation.

a. $y = 3^x$ **b.** $y = 0.3^x$

◆ Solution
a. Because the base 3 is greater than 1, the effects of x can be seen in a chart.

x	-3	-2	-1	0	1	2	3	4
$y = 3^x$	$\frac{1}{27}$	$\frac{1}{9}$	$\frac{1}{3}$	1	3	9	27	81

From the chart, it is apparent that as x increases from zero, the function value gets larger. At $x = 0$, y is equal to 1 and, as x decreases from zero, the values of the function get smaller and approach zero.

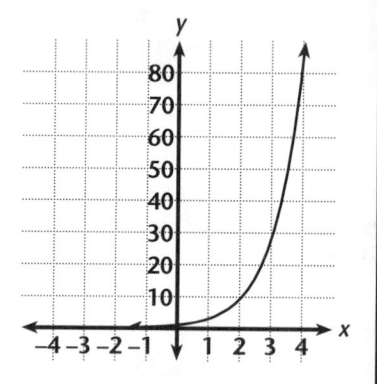

b. Because the base is less than 1, the effects of x will be different from the previous example.

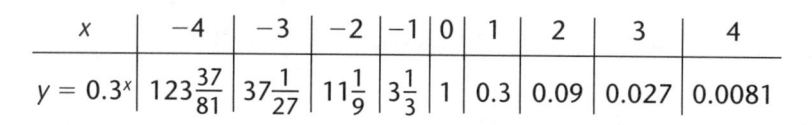

x	-4	-3	-2	-1	0	1	2	3	4
$y = 0.3^x$	$123\frac{37}{81}$	$37\frac{1}{27}$	$11\frac{1}{9}$	$3\frac{1}{3}$	1	0.3	0.09	0.027	0.0081

From this chart, it is obvious that as x increases from zero, the function values decrease. At $x = 0$, the y value is equal to 1 and, as x decreases from zero, the values of the function get larger.

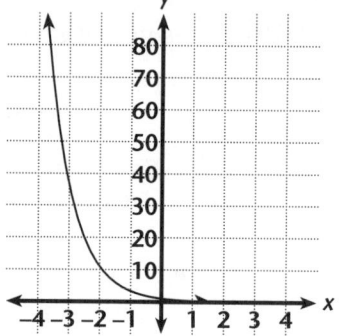

For each function, describe the effect on y as x increases from negative to positive. Sketch the graph on graph paper.

6. $y = 5^x$ _____ **7.** $y = 0.5^x$ _____

8. $y = 8^x$ _____ **9.** $y = 0.8^x$ _____

Reteaching

8.7 Applications of Exponential Functions

◆ **Skill A** Using exponential functions to model applications

Recall An exponential function results when a value increases by a constant multiplier.

◆ **Example**
Model the following problem as an exponential function:
   A stock has been increasing in value by 5% each year since its purchase 4 years ago. If it was purchased at $62, what is its present value?

◆ **Solution**
If the initial value of the stock was $62, it increased in value by 5% of $62 after the first year, or $62 + 62 \cdot 0.05$. By the Distributive Property, this expression can be rewritten as $62(1 + 0.5)$. The factor $(1 + 0.05)$ is the multiplier, which increases the new value by 0.05, or 5%. Therefore, the value of the stock after 4 years can be written as follows:

$$62 \cdot (1 + 0.05) \cdot (1 + 0.05) \cdot (1 + 0.05) \cdot (1 + 0.05)$$

↑ initial value ↑ value after year 1 ↑ value after year 2 ↑ value after year 3 ↑ value after year 4

The product above can also be written in as $62 \cdot (1 + 0.05)^4$. Notice that this product has the form of the General Growth Formula, $P = A(1 + r)^t$. In this situation, P represents the present value, A represents the initial investment, r represents the annual rate of increase, and t is the number of years of investment. Thus the original problem can be modeled by the exponential equation $P = 62 \cdot (1 + 0.05)^4$.

Model each problem as an exponential function.

1. A baseball card that originally cost $24 has been decreasing in value at an annual rate of 6% for the last 5 years. What is its present value?

2. A large city has a population of 500,000 and has been increasing in size at an annual rate of 2% for 6 years. What was its population 6 years ago?

3. One share of stock that originally sold for $54 has increased in value at an annual rate of 4% for the last 7 years. What is its present value?

Algebra 1 Reteaching Masters **97**

◆ **Skill B** Using exponential functions to solve problems about growth and decay

Recall When an expression contains more than one operation, including parentheses and exponents, the expression within parentheses is simplified first.

◆ **Example**
If $2000 is deposited in a savings account paying 7% interest compounded quarterly, what will the account be worth in 8 years if no deposits or withdrawals are made? (Use the compound interest formula, $A = P\left(1 + \frac{r}{n}\right)^{nt}$, where A is the amount after t years, P is the principal invested, r is the annual interest rate in decimal form, and n is the number of times that the interest is compounded per year.)

◆ **Solution**
Substituting the given values into the formula, the equation becomes $A = 2000\left(1 + \frac{0.07}{4}\right)^{4 \times 8}$. Recognize that compounding quarterly means that the interest is applied 4 times per year, so $n = 4$ in this example. Simplifying $\frac{0.07}{4}$, it becomes 0.0175. Thus, $A = 2000(1 + 0.0175)^{4 \times 8}$, or $A = 2000(1.0175)^{32}$. Next, the exponential factor is evaluated. This power is approximately 1.7422. By multiplying this result and rounding to the nearest cent, you get $A = 3484.40$. In 8 years, the account will be worth $3484.40.

Solve each problem.

4. If $5000 is invested in an account paying 6% interest compounded semiannually, how much will the investment be worth in 10 years? $\left(\text{Use } A = P\left(1 + \frac{r}{n}\right)^{nt}.\right)$

5. A bacterial culture grows according to the formula $N = S(2.54)^{0.04t}$, where S is the starting number of bacteria, and N is the number of bacteria after T hours. How many bacteria will be present after 24 hours if the initial number is 12,000?

6. The population of a certain town decreases according to the formula $P = 8000(0.9)^{0.2t}$, where t is the number of years after 1990. What will the population be in the year 2000?

Reteaching
9.1 Adding and Subtracting Polynomials

◆ **Skill A** Adding polynomials

Recall To add two polynomials, add the coefficients of like terms.

◆ **Example 1**
Add the polynomials horizontally.
$3a^3 + 2a^2 + a + 5$ and $2a^3 + 4a - 6$

◆ **Solution**
Group like terms.
$(3a^3 + 2a^2 + a + 5) + (2a^3 + 4a - 6)$
$= (3a^3 + 2a^3) + 2a^2 + (a + 4a) + (5 - 6)$
$= 5a^3 + 2a^2 + 5a - 1$
The sum of $3a^3 + 2a^2 + a + 5$ and $2a^3 + 4a - 6$ is $5a^3 + 2a^2 + 5a - 1$.

◆ **Example 2**
Add the same two polynomials vertically.

◆ **Solution**
Line up the variables. Use zero for the coefficient of any missing variable.
$$\begin{array}{r} 3a^3 + 2a^2 + 1a + 5 \\ + \; 2a^3 + 0a^2 + 4a - 6 \\ \hline 5a^3 + 2a^2 + 5a - 1 \end{array}$$
The sum of $3a^3 + 2a^2 + a + 5$ and $2a^3 + 4a - 6$ is $5a^3 + 2a^2 + 5a - 1$.

Find each sum.

1. $(5b^2 + 3b) + (b^2 - 2b)$ _____

2. $(8c^2 - 2c) + (2c^2 + 3c)$ _____

3. $(b^3 + 2b^2 + 3b) + (4b^3 - 5b^2 + 4b)$ _____

4. $(3y^3 + 3y - 1) + (2y^3 + 5y^2 + 3y)$ _____

5. $(5r^2 + 3r + 6) + (2r^3 + r^2 + 4r)$ _____

6. $(4m^3 - 5m^2 - m) + (3m^3 - 3m - 5)$ _____

7. $(2x^2 + 3x + 4) + (-5x^2 + x - 7)$ _____

8. $(x^2 - x + 6) + (3x^2 - x + 3)$ _____

9. $(2x^2 + 3x + 6) + (-2x^2 - 7)$ _____

10. $(4x^3 - 5x + 4) + (3x^3 + 5x - 3)$ _____

Algebra 1   Reteaching Masters   **99**

NAME _____ CLASS _____ DATE _____

◆ **Skill B** Finding the opposite of a polynomial

Recall To find the opposite of a term, change the sign in front of the term.

◆ **Example**
Find the opposite of $2b^2 + 3b - 7$.

◆ **Solution**
The opposite of $2b^2 + 3b - 7$ is $-(2b^2 + 3b - 7)$.
$-(2b^2) = 2b^2$; $-(3b) = -3b$; $-(-7) = 7$
Thus, $-(2b^2 + 3b - 7) = -2b^2 - 3b + 7$.

Find the opposite of each polynomial.

11. $3c^2 + c + 5$ _____ 12. $n^2 - 2n + 3$ _____

13. $-2z^2 - z - 1$ _____ 14. $5r^2 + 4r - 9$ _____

15. $4t^2 - t$ _____ 16. $-9q^2 - q - 3$ _____

17. $5 - 2a - 3a^2$ _____ 18. $5e^3 - 4e^2 + 2e$ _____

◆ **Skill C** Subtracting polynomials

Recall To subtract a polynomial, add its opposite.

◆ **Example**
Subtract $2c^2 - 3c - 5$ from $5c^2 - 2c + 3$.

◆ **Solution**
$(5c^2 - 2c + 3) - (2c^2 - 3c - 5) = (5c^2 - 2c + 3) + (-2c^2 + 3c + 5)$
$= 3c^2 + c + 8$
$2c^2 - 3c - 5$ subtracted from $5c^2 - 2c + 3$ is $3c^2 + c + 8$.

Find each difference.

19. $(x^2 + 3x + 2) - (3x^2 + x - 6)$ _____

20. $(2x^2 - 5x + 1) - (2x^2 + 3x - 2)$ _____

21. $(3x^2 + x - 4) - (-3x^2 + 2x - 7)$ _____

22. $(-x^2 - x - 4) - (-2x^2 - 4x + 3)$ _____

Simplify. Express all answers in standard form.

23. $(2x^2 + 3x - 4) - (2x - 5) + (x^2 - x + 1)$ _____

24. $(-5x^2 - 2x + 1) - (3x^2 + 4x - 2) - (-8x^2 - 5x - 3)$ _____

100   Reteaching Masters   Algebra 1

Reteaching
9.2 Modeling Polynomial Multiplication

◆**Skill A** Using models to multiply monomials and binomials

Recall Algebra tiles can be used to find products, such as that of 3 and $x + 1$. Three positive 1-tiles are placed along the left line. One positive x-tile and one positive 1-tile are placed along the top line. Place tiles that form a rectangular shape inside the lines.

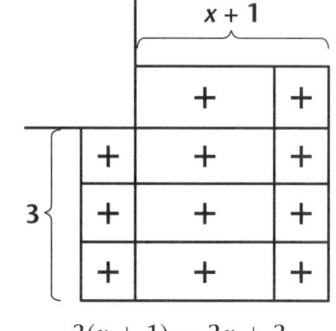

$3(x + 1) = 3x + 3$

◆ **Example 1**
Use algebra tiles to find the product $2(x + 2)$.

◆ **Solution**

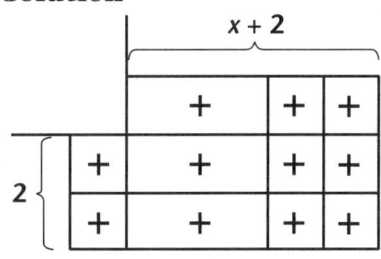

Thus, $2(x + 2) = 2x + 4$.

◆ **Example 2**
Use algebra tiles and the Distributive Property to find the product $x(x - 1)$.

◆ **Solution**

Using Tiles

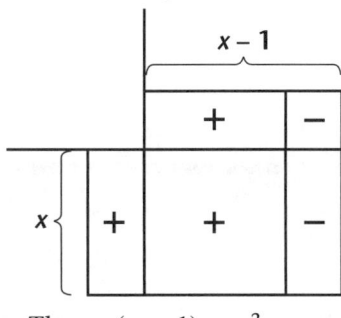

Using the Distributive Property
$x(x - 1) = (x \cdot x) - (x \cdot 1)$
$= x^2 - 1$

Thus, $x(x - 1) = x^2 - x$.

Find each product.

1. $2(x + 3)$ _____ 2. $3(2x - 2)$ _____

3. $x(x - 3)$ _____ 4. $x(-x + 1)$ _____

5. $x(2x + 4)$ _____ 6. $x(3x - 3)$ _____

7. $2x(x - 2)$ _____ 8. $2x(3x + 2)$ _____

Algebra 1   Reteaching Masters   **101**

◆ **Skill B** Using models to multiply two binomials

Recall Algebra tiles can also be used to find the product of two binomials.

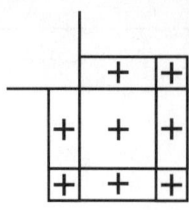

This model shows that $(x + 1)(x + 1) = x^2 + 2x + 1$.

◆ **Example 1**
Use algebra tiles to find the product $(x - 2)(x - 3)$.

◆ **Solution**

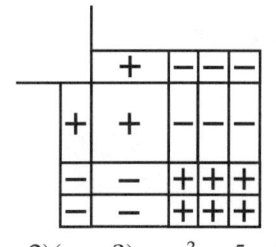

$(x - 2)(x - 3) = x^2 - 5x + 6$

◆ **Example 2**
Use algebra tiles to find the product $(x + 2)(x - 2)$.

◆ **Solution**

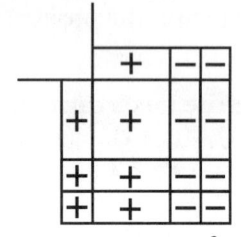

$(x + 2)(x - 2) = x^2 - 4$

Use algebra tiles to find each product.

9. $(x + 2)(x + 1)$ _____   10. $(x + 2)(x - 1)$ _____

11. $(x - 2)(x + 1)$ _____   12. $(x - 2)(x - 1)$ _____

13. $(x + 3)(x + 3)$ _____   14. $(x - 3)(x - 3)$ _____

15. $(x + 3)(x - 3)$ _____   16. $(x - 3)(x + 3)$ _____

17. $(x + 1)(2x + 3)$ _____   18. $(x - 2)(2x - 1)$ _____

19. $(3x + 2)(x - 2)$ _____   20. $(2x + 1)(3x - 2)$ _____

NAME _____ CLASS _____ DATE _____

Reteaching
9.3 Multiplying Binomials

◆ **Skill A** Multiplying monomials and binomials

Recall The Distributive Property can be used to find the product of a monomial and a binomial.

◆ **Example**
Use the Distributive Property to find the product $x(x - 4)$.

◆ **Solution**
$x(x - 4) = x(x) - x(4)$
$= x^2 - 4x$
The product $x(x - 4)$ is $x^2 - 4x$.

Use the Distributive Property to find each product.

1. $4(x + 5)$ _____

2. $5(x - 2)$ _____

3. $x(2x - 2)$ _____

4. $2x(3x + 1)$ _____

5. $-5x(x - 6)$ _____

6. $-3x(-x - 3)$ _____

◆ **Skill B** Multiplying two binomials

Recall The Distributive Property can also be used to multiply two binomials.

◆ **Example**
Use the Distributive Property to find the product $(x + 2)(x + 5)$.

◆ **Solution**
$(x + 2)(x - 5) = (x + 2)(x) - (x + 2)(5)$
$= x(x) + (2)(x) - (x)(5) - (2)(5)$
$= x^2 + 2x - 5x - 10$
$= x^2 - 3x - 10$

Use the Distributive Property to find each product.

7. $(x + 1)(x + 4)$

8. $(x + 3)(x + 2)$

9. $(x + 5)(x - 3)$

10. $(2x + 3)(x + 2)$

11. $(x - 5)(3x - 3)$

12. $(3x - 4)(4x - 3)$

Algebra 1   Reteaching Masters   **103**

NAME _____ CLASS _____ DATE _____

◆ **Skill C** Using the FOIL method to multiply two binomials

Recall To multiply two binomials,
multiply the **F**irst terms;
multiply the **O**utside terms;
multiply the **I**nside terms;
add the outside and inside products; and
multiply the **L**ast terms.

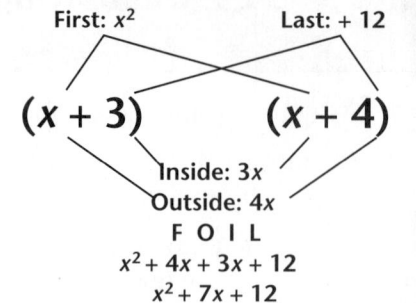

First: x^2 Last: $+12$

$(x + 3)\quad(x + 4)$

Inside: $3x$
Outside: $4x$

F O I L
$x^2 + 4x + 3x + 12$
$x^2 + 7x + 12$

◆ **Example 1**
Use the FOIL method to find the product
$(x + 3)(x - 4)$.

◆ **Solution**

$$\begin{array}{r} \quad\text{F}\quad\quad\text{O}\quad\quad\text{I}\quad\quad\text{L} \\ (x + 3)(x - 4) = (x)(x) + (x)(-4) + (3)(x) + (3)(-4) \\ = x^2 \quad - 4x \quad + 3x \quad - 12 \\ = x^2 - x - 12 \end{array}$$

The product $(x + 3)(x - 4)$ is $x^2 - x - 12$.

◆ **Example 2**
Use the FOIL method to find the product $(2x - 3)(3x - 1)$.

◆ **Solution**

$$\begin{array}{r} \quad\text{F}\quad\quad\text{O}\quad\quad\text{I}\quad\quad\text{L} \\ (2x - 3)(3x - 1) = (2x)(3x) + (2x)(-1) + (-3)(3x) + (-3)(-1) \\ = 6x^2 - 2x - 9x + 3 \\ = 6x^2 - 11x + 3 \end{array}$$

The product $(2x - 3)(3x - 1)$ is $6x^2 - 11x + 3$.

Use the FOIL method to find each product.

13. $(x + 2)(x + 5)$

14. $(x + 3)(x - 4)$

15. $(x - 5)(x - 3)$

16. $(2x - 3)(x + 3)$

17. $(2x - 5)(2x - 5)$

18. $(5x - 4)(5x - 4)$

19. $(4x + 1)(x - 1)$

20. $(2x - 3)(x - 2)$

21. $(-3x + 4)(2x - 3)$

104 Reteaching Masters   Algebra 1

NAME _____ CLASS _____ DATE _____

Reteaching
9.4 Polynomial Functions

◆ **Skill A** Evaluating a polynomial function

Recall To evaluate a polynomial function, substitute the value of the variable. Then simplify.

◆ **Example**
Given $x = -3$, evaluate the function $y = x^2 + 3x + 2$.

◆ **Solution**
Substitute -3 for x.
$y = (-3)^2 + 3(-3) + 2$
$= 9 - 9 + 2$
$= 2$
When $x = -3$, $y = 2$.

Given $x = 4$, evaluate each function.

1. $y = x^2 - x - 12$ _____   2. $y = 2x^2 + 5x + 2$ _____

Given $m = -2$ and $n = 3$, evaluate each function.

3. $A = m^2n$ _____   4. $B = m^2 + 2mn + n^2$ _____

◆ **Skill B** Evaluating volume and surface area functions

Recall The volume of a box with a square base of length x and height 10 inches can be represented by the function $V = 10x^2$.
The surface area of the same box can be represented by the function $S = 2x^2 + 40x$.

◆ **Example**
Find the volume and surface area of a box with a square base if the side length of the square is 3 inches and the height of the box is 10 inches.

◆ **Solution**
To find the volume, substitute 3 for x in the volume function.
$V = 10(3)^2$
$= (10)(9)$
$= 90$
Thus, the volume of the box is 90 cubic inches.
To find the surface area, substitute 3 for x in the surface area function.
$S = 2(3)^2 + 40(3)$
$= 18 + 120$
$= 138$
Thus, the surface area of the box is 138 square inches.

Algebra 1   Reteaching Masters   **105**

NAME _____ CLASS _____ DATE _____

A box has a square base of 5 meters on each side. If the height is 3 meters, the volume of the box is given by V = $3x^2$, and the surface area is given by S = $2x^2 + 12x$. Find the volume and surface area.

5. volume _____ 6. surface area _____

A box has a square base of 2.5 feet on each side. If the height is 20 feet, the volume of the box is given by V = $20x^2$, and the surface area is given by S = $2x^2 + 80x$. Find the volume and the surface area.

7. volume _____ 8. surface area _____

◆ **Skill C** Showing that an equation is an identity

Recall An equation in which the expressions on each side of the equal sign are equivalent expressions is called an identity.

◆ **Example**
Show that the equation $(x + 1)(x - 2) = x^2 - x - 2$ is an identity by substituting the integers from −2 to 2 inclusive for x.

◆ **Solution**
Use substitution to make a table of values for $y_1 = (x + 1)(x - 2)$ and $y_2 = x^2 - x - 2$.

x	−2	−1	0	1	2
y_1	4	0	−2	−2	0
y_2	4	0	−2	−2	0

Show that each equation is an identity by substituting integers from −2 to 2 for x.

9. $(x + 2)(x + 1) = x^2 + 3x + 2$

x	−2	−1	0	1	2
y_1					
y_2					

10. $x^2 - x - 6 = (x - 3)(x + 2)$

x	−2	−1	0	1	2
y_1					
y_2					

11. $(x - 5)(x - 2) = x^2 - 7x + 10$

x	−2	−1	0	1	2
y_1					
y_2					

12. $(x + 4)(x + 4) = x^2 + 8x + 16$

x	−2	−1	0	1	2
y_1					
y_2					

13. $x^2 - 4 = (x + 2)(x - 2)$

x	−2	−1	0	1	2
y_1					
y_2					

14. $(x - 1)(x - 1) = x^2 - 2x + 1$

x	−2	−1	0	1	2
y_1					
y_2					

Reteaching Masters Algebra 1

NAME _____ CLASS _____ DATE _____

Reteaching
9.5 Common Factors

◆ Skill A Listing prime numbers

Recall A positive integer is a prime number if it has exactly two factors: 1 and itself. The number 3 is prime because it has exactly two factors: 1 and 3. The number 4 is not prime because it has three factors: 1, 2, and 4.

- **◆ Example**
  Which of the following whole numbers are prime?
  a. 1   **b.** 37   **c.** 87

- **◆ Solution**
  a. 1 is not prime because it has only one factor, itself.
  b. 37 is prime because it has exactly two factors: 1 and 37.
  c. 87 is not prime because it has more than two factors: 1, 3, 29, and 87.

List the factors of each whole number. Tell whether each number is prime.

1. 12 _____
2. 35 _____
3. 47 _____

4. 57 _____
5. 77 _____
6. 97 _____

◆ Skill B Factoring out a common monomial

Recall The product of x and $x + 3$ is $x^2 + 3x$. To factor an expression such as $x^2 + 3x$ means to factor out the greatest common factor (GCF) of each term. The Distributive Property is used to rewrite the expression as the product $x(x + 3)$. Some expressions cannot be factored. The greatest common factor of $x^2 - 3$ is 1, so $x^2 - 3$ is considered prime.

- **◆ Example**
  Factor each polynomial.
  a. $4b^3 - 6b^2 + 10b$ **b.** $5y^3 - 15x^2 + 18xy$

- **◆ Solution**
  a. The GCF is $2b$.
  $4b^3 - 6b^2 + 10b = 2b(2b^2 - 3b + 5)$
  b. The GCF is 1.
  $5y^3 - 15x^2 + 18xy$ is prime.

Factor each polynomial.

7. $3m^2 - 21m$ _____
8. $8t^2 + 15t$ _____

9. $18p^2 + 21p + 9$ _____
10. $4d^3 - 20d^2 + 8d$ _____

Algebra 1   Reteaching Masters   **107**

NAME _____ CLASS _____ DATE _____

◆ **Skill C** Factoring out a common binomial factor

Recall When an expression has a common binomial factor, the Distributive Property is used to rewrite the expression as a product. Sometimes the expression must be regrouped first.

◆ **Example 1**
Factor $a(b - 1) + 3(b - 1)$.

◆ **Solution**
The common binomial factor is $b - 1$.
$a(b - 1) + 3(b - 1) = (b - 1)(a + 3)$

◆ **Example 2**
Factor $x^2 - 3x + 6x - 18$.

◆ **Solution**
Group the first two terms together. Group the last two terms together.
$(x^2 - 3x) + (6x - 18)$
Use the Distributive Property to factor the grouped terms.
$x(x - 3) + 6(x - 3)$
Factor out the common binomial.
$(x - 3)(x + 6)$
$x^2 - 3x + 6x - 18 = (x + 6)(x - 3)$

◆ **Example 3**
Factor $x^2 + 2x - 4x - 8$.

◆ **Solution**
$x^2 + 2x - 4x - 8 = (x^2 + 2x) - (4x + 8)$
$= x(x + 2) - 4(x + 2)$
$= (x + 2)(x - 4)$

Factor each polynomial.

11. $x(x + 5) + 8(x + 5)$

12. $(x - 7)x - (x - 7)3$

13. $x^2 + 3x + 4x + 12$

14. $x^2 - 3x - 2x + 6$

15. $x^2 + 5x - 2x - 10$

16. $x^2 + x - 4x - 4$

108   Reteaching Masters   Algebra 1

NAME _____ CLASS _____ DATE _____

Reteaching
9.6 Factoring Special Polynomials

◆ **Skill A** Factoring a perfect-square binomial

Recall The binomial $(x - y)^2$ can be written as a trinomial by using the FOIL method.
$(x - y)^2 = (x - y)(x - y)$
$= x^2 - xy - xy + y^2$
$= x^2 - 2xy + y^2$

When a binomial is squared, the product has a pattern.
  The first term is one variable squared, x^2.
  The middle term is twice the product of the two variables, $2xy$.
  The second term is the second variable squared, y^2.
  If the binomial is a difference, the middle term is subtracted. If the binomial is a sum, the middle term is added.
The expression $x^2 + 2xy + y^2$ is called a perfect-square trinomial because it is written in the form $(x + y)^2$ when it is factored. If a trinomial has a perfect-square trinomial pattern, it can be factored by reversing the pattern.

◆ **Example**
Factor each expression.

   a. $s^2 + 6s + 9$ **b.** $4m^2 - 12mn + 9n^2$

◆ **Solution**
   a. The first term is a perfect square, s^2.
   The last term is a perfect square, 3^2.
   The middle term, $6s$, is $(2)(s)(3)$.
   $s^2 + 6s + 9 = (s + 3)(s + 3)$
   $\qquad\qquad\quad = (s + 3)^2$

   b. The first term is a perfect square, $(2m)^2$.
   The last term is a perfect square, $(3n)^2$.
   The middle term, $12mn$, is $(2)(2m)(3n)$.
   $4m^2 - 12mn + 9n^2 = (2m - 3n)(2m - 3n)$
   $\qquad\qquad\qquad\qquad\;\; = (2m - 3n)^2$

Square each expression in parentheses.

1. $(x + 4)^2$ _____ 2. $(2v - 5)^2$ _____

3. $(3d - a)^2$ _____ 4. $(10k - t)^2$ _____

Factor each polynomial completely. If the polynomial cannot be factored, write *prime*.

5. $x^2 + 2x + 1$ _____ 6. $y^2 - 10y - 25$ _____

7. $4p^2 + 16p + 16$ _____ 8. $4a^2 - 4ab + b^2$ _____

9. $r^2 - 4rs + 4s^2$ _____ 10. $49b^2 - 42bc + 9c^2$ _____

Algebra 1 Reteaching Masters 109

NAME _____ CLASS _____ DATE _____

◆ Skill B Factoring the difference of two squares

Recall The product $(x + y)(x - y)$ can be written as a binomial by using the FOIL method.
$$(x + y)(x - y) = x^2 - xy + xy - y^2$$
$$= x^2 - y^2$$

When the sum and difference of two monomials are multiplied, the product has a pattern:
- The first term is one variable squared, x^2.
- The second term is the other variable squared, y^2.
- The middle term is 0.

An expression written in the form $x^2 - y^2$ is called the difference of two squares. When a binomial is written as the difference of two squares, the binomial can be factored into the sum and difference of two monomials.

◆ **Example**
Factor each expression.

a. $r^2 - 16$ **b.** $4b^2 - 9c^2$ **c.** $n^4 - 4$

◆ **Solution**

a. The first term is a perfect square, r^2.
The last term is a perfect square, 4^2.
$$r^2 - 16 = r^2 - 4^2$$
$$= (r + 4)(r - 4)$$

b. The first term is a perfect square, $(2b)^2$.
The last term is a perfect square, $(3c)^2$.
$$4b^2 - 9c^2 = (2b)^2 - (3c)^2$$
$$= (2b + 3c)(2b - 3c)$$

c. The first term is a perfect square, $(n^2)^2$.
The last term is a perfect square, 2^2.
$$n^4 - 4 = (n^2)^2 - 2^2$$
$$= (n^2 + 2)(n^2 - 2)$$

Find each product.

11. $(t + 5)(t - 5)$ _____   **12.** $(m - d)(m + d)$ _____

13. $(3r - s^2)(3r + s^2)$ _____   **14.** $(5p + 4q)(5p - 4q)$ _____

Factor each polynomial completely. If the polynomial cannot be factored, write *prime*.

15. $t^2 - 49$ _____   **16.** $4 - a^2$ _____

17. $9d^2 + 1$ _____   **18.** $4s^2 - t^2$ _____

19. $25c^2 - 4q^2$ _____   **20.** $16b^2 - c^4$ _____

21. $m^2n^2 - p^2$ _____   **22.** $s^4 - t^4$ _____

Reteaching
9.7 Factoring Quadratic Trinomials

◆ **Skill A** Factoring trinomials by using diagrams

Recall The product of $a + 2$ and $a - 5$ can be found by using the FOIL method.
$(a + 2)(a - 5) = a^2 - 5a + 2a - 10$
$\qquad\qquad\qquad = a^2 - 3a - 10$
Factoring a trinomial, such as $a^2 - 3a - 10$, means writing the expression as a product of monomials or binomials whose greatest common factor is one. If no monomials or binomials can be found, the trinomial is prime.

◆ **Example**
Use a diagram to factor $x^2 - 6x + 8$.

◆ **Solution**
First fill the boxes with information from the trinomial.

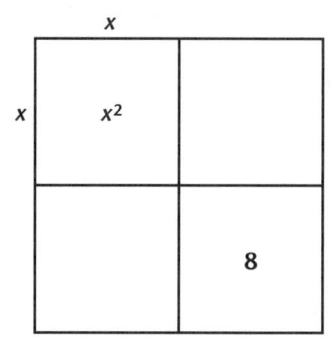

	x	
x	x^2	
		8

Complete the box with the two factors that have a product of 8 and a sum of -6.

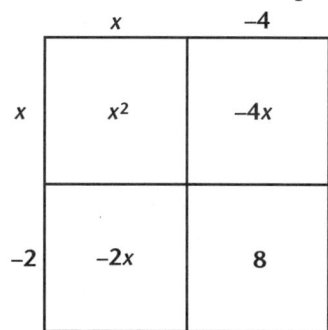

	x	-4
x	x^2	$-4x$
-2	$-2x$	8

Thus, $x^2 - 6x + 8 = (x - 2)(x - 4)$.

Use a diagram to factor each trinomial.

1. $x^2 + 3x + 2$
2. $x^2 + x - 12$
3. $x^2 - 10x + 21$

4. $x^2 + 4x - 5$
5. $x^2 + 11x + 24$
6. $x^2 - 8x + 16$

Algebra 1 — Reteaching Masters 111

NAME _____ CLASS _____ DATE _____

◆ **Skill B** Factoring trinomials by choosing factor pairs of the constant

Recall Another way to factor a trinomial, such as $x^2 - 5x - 6$, is to first make a list of the pairs of factors of the constant. Then choose the right combination to complete the factors of the trinomial.

◆ **Example**
Use the constant's factor pairs to factor $x^2 - 5x - 6$.

◆ **Solution**
List each pair of factors of -6 along with their sum.

Factors of -6	Sum of the factors
6 and -1	5
3 and -2	1
2 and -3	-1
1 and -6	-5

The sum of 1 and -6 is -5. Use the combination of 1 and -6 to form the factors. Thus, $x^2 - 5x - 6 = (x + 1)(x - 6)$.

Factor each trinomial. If the trinomial cannot be factored, write *prime*.

7. $x^2 - x - 2$

8. $x^2 + 3x - 4$

9. $x^2 + 4x + 3$

10. $x^2 - 4x + 3$

11. $x^2 + 2x - 8$

12. $x^2 + x - 20$

13. $x^2 + 2x - 15$

14. $x^2 - 3x + 10$

15. $x^2 - x - 12$

16. $x^2 + 6x + 8$

17. $x^2 - 20x + 36$

18. $x^2 + 2x - 24$

112   Reteaching Masters   Algebra 1

NAME _____ CLASS _____ DATE _____

Reteaching
9.8 Solving Equations by Factoring

◆ **Skill A** Using the Zero-Product Property

Recall If a and b are real numbers such that $ab = 0$, then $a = 0$ or $b = 0$.

◆ **Example**
Solve the equation $(x + 3)(x - 1) = 0$.

◆ **Solution**
Use the Zero-Product Property. If the product of two factors is equal to 0, then one of the factors must be 0. Set each factor equal to 0 and solve.

First factor   Second factor
$(x + 3) = 0$ $(x - 1) = 0$
  $x = -3$ $x = 1$

Check by substituting in the original equation.
Substitute -3 for x: $(-3 + 3)(-3 - 1) = 0(-4) = 0$
Substitute 1 for x: $(1 + 3)(1 - 1) = 4(0) = 0$

The equation $(x + 3)(x - 1) = 0$ has two solutions. The solutions are -3 and 1.

Solve by factoring.

1. $(x - 3)(x - 2) = 0$

2. $(x + 5)(x - 4) = 0$

3. $(x + 4)(x - 4) = 0$

4. $(x - 6)(x - 6) = 0$

5. $(x - 2.8)(x + 5.2) = 0$

6. $\left(x + \frac{2}{3}\right)(x - 1) = 0$

7. $(2x - 4)(3x - 6) = 0$

8. $(5x + 3)(4x + 7) = 0$

9. $\left(\frac{3}{5}x + 6\right)\left(\frac{7}{8}x - 1\right) = 0$

10. $(4.7x + 14.1)(2.4x - 3.6) = 0$

Algebra 1   Reteaching Masters

NAME _____ CLASS _____ DATE _____

◆ **Skill B** Find the zeros of a polynomial function by factoring

Recall The zeros of a function are the values of x that make y equal to 0.

◆ **Example 1**
Find the zeros of the function $y = (x - 2)(x + 5)$.

◆ **Solution**
Let $y = 0$. Then use the Zero-Product Property to solve for x.

$(x - 2)(x + 5) = 0$
$(x - 2) = 0$ or $(x + 5) = 0$
$x = 2$ or $x = -5$
The zeros of $y = (x - 2)(x + 5)$ are 2 and -5.

Recall A quadratic polynomial can be factored into two binomials.

◆ **Example 2**
Solve the equation $x^2 - x - 6 = 0$.

◆ **Solution**
Since $x^2 - x - 6$ can be factored into $(x + 2)(x - 3)$, you can rewrite $x^2 - x - 6 = 0$ as $(x + 2)(x - 3) = 0$. Solve the equation $(x + 2)(x - 3) = 0$.
$x + 2 = 0$ or $x - 3 = 0$
$x = -2$ or $x = 3$
The solutions to $x^2 - x - 6 = 0$ are -2 and 3.

Solve by factoring.

11. $x^2 - 4x - 12 = 0$

12. $x^2 - 6x + 9 = 0$

13. $x^2 - 9x + 14 = 0$

14. $x^2 + 6x + 5 = 0$

15. $x^2 - 7x + 10 = 0$

16. $x^2 - 36 = 0$

17. $x^2 + 8x + 16 = 0$

18. $x^2 - x - 12 = 0$

19. $9x^2 - 1 = 0$

20. $4x^2 + 4x + 1 = 0$

114 Reteaching Masters Algebra 1

NAME _____ CLASS _____ DATE _____

Reteaching
10.1 Graphing Parabolas

◆ **Skill A**   Finding the vertices and axes of symmetry of parabolas

Recall   A quadratic function in the form $y = a(x - h)^2 + k$, where $a \neq 0$, transforms the parent function $y = x^2$ by
- stretching the parent function by a factor of a,
- moving the vertex of the parent function from (0, 0) to (h, k), and
- moving the axis of symmetry to $x = h$.

◆ **Example**
Identify the vertex and axis of symmetry of $y = 3(x - 2)^2 - 4$. Then sketch the graph.

◆ **Solution**
The value for h is 2.
The value for k is -4.
The value for a is 3.
Thus, the vertex is $(2, -4)$, and the axis of symmetry is $x = 2$.

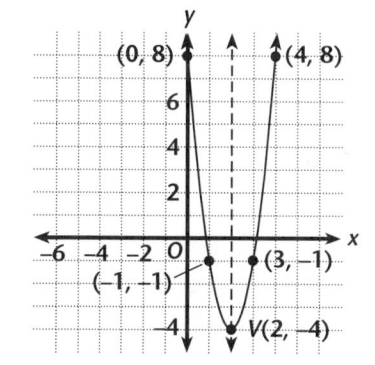

Identify the vertex and axis of symmetry for each quadratic function. Then sketch the graph.

1. $y = 2(x - 5)^2 + 1$

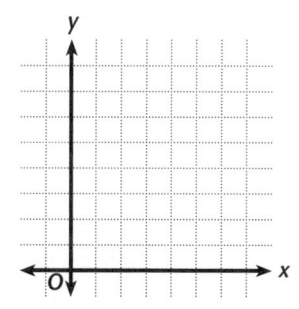

2. $y = -(x - 4)^2 - 2$

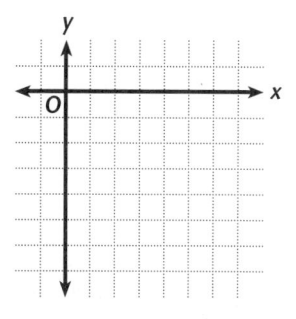

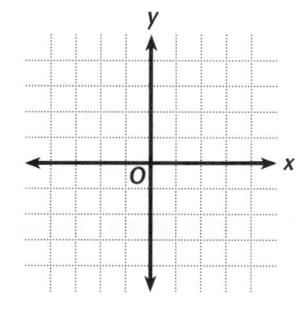

3. $y = -2(x + 1)^2 + 4$

4. $y = (x - 3)^2 + 5$

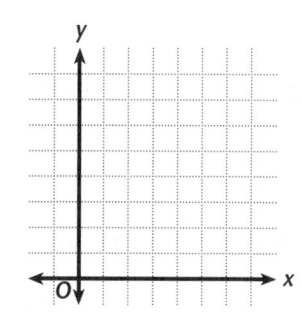

Algebra 1   Reteaching Masters **115**

NAME _____ CLASS _____ DATE _____

◆ **Skill B** Finding the zeros of quadratic functions

Recall The zeros of a quadratic function in the form $y = ax^2 + bx + c$ are the x-values for which y is equal to 0. When a sketch of the function is drawn, the zeros of the function are the x-values of the points where the parabola intersects the x-axis.

◆ **Example**
Find the zeros of $y = x^2 - x - 6$ by graphing.

◆ **Solution**

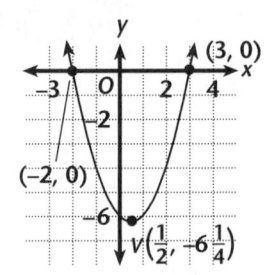

The zeros are 3 and −2.

Sketch the graph of each quadratic function. Then find its zeros.

5. $y = x^2 - 3x + 2$ _____

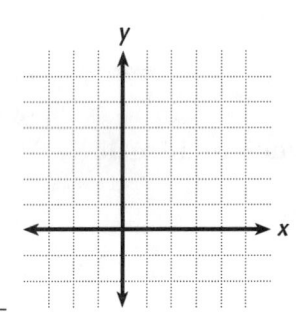

6. $y = x^2 - 3x - 10$ _____

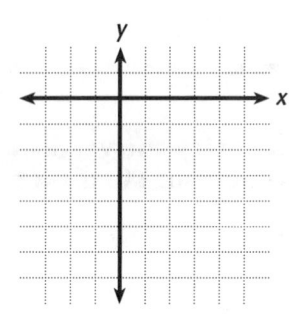

◆ **Skill C** Finding the zeros of quadratic functions by using the Zero-Product Property

Recall The zeros of a quadractic function are the values of x for which y is equal to 0. Thus, you can also find the zeros by factoring and then using the Zero-Product Property.

◆ **Example**
Find the zeros of $y = x^2 - x - 6$ by factoring.

◆ **Solution**
Let $y = 0$.
$x^2 - x - 6 = 0$
$(x - 3)(x + 2) = 0$
$x = 3$ or $x = -2$
The zeros of $y = x^2 - x - 6$ are 3 and −2.

Use factoring to find the zeros of each function.

7. $y = x^2 + 4x + 3$ _____

8. $y = x^2 + 2x + 2$ _____

9. $y = 2x^2 - 5x + 2$ _____

10. $y = 2x^2 + x - 6$ _____

NAME _____ CLASS _____ DATE _____

Reteaching
10.2 Solving Equations by Using Square Roots

◆ **Skill A** Finding square roots

Recall Every positive number has a positive and a negative square root. For example, the positive square root of 4 is 2 because 2 · 2 = 4. The negative square root of 4 is −2 because −2 · −2 = 4. The positive root of 4 is indicated by $\sqrt{4}$. The negative square root of 4 is indicated by $-\sqrt{4}$. Thus, $\sqrt{4} = 2$ and $-\sqrt{4} = -2$.

◆ **Example**
Find each square root.

a. $\sqrt{169}$ b. $-\sqrt{64}$ c. $\sqrt{45}$

◆ **Solution**

a. $\sqrt{169} = 13$ b. $-\sqrt{64} = -8$ c. $\sqrt{45} \approx 6.71$

Find each positive square root. Round answers to the nearest hundredth when necessary.

1. $\sqrt{36}$ _____   2. $-\sqrt{196}$ _____   3. $-\sqrt{275}$ _____

4. $\sqrt{484}$ _____   5. $\sqrt{240}$ _____   6. $-\sqrt{841}$ _____

◆ **Skill B** Solving equations of the form $x^2 = k$, where $k \geq 0$

Recall There are two solutions to the equation $x^2 = 4$: $\sqrt{4} = 2$ and $-\sqrt{4} = -2$. If $x^2 = 4$, x is equal to 2 or −2; that is, $x = \pm 2$. The solutions are 2 and −2.

◆ **Example**
Solve each equation. Round to the nearest hundredth, if necessary.

a. $x^2 = 49$ b. $x^2 = 115$ c. $x^2 = 6.25$

◆ **Solution**

a. $x = \pm 7$; the solutions are 7 and −7.
b. $x = \pm 10.72$; the solutions are approximately 10.72 and −10.72.
c. $x = \pm 2.5$; the solutions are 2.5 and −2.5.

Solve each equation. Round answers to the nearest hundredth when necessary.

7. $x^2 = 25$ _____   8. $x^2 = 75$ _____   9. $x^2 = 108$ _____

Algebra 1   Reteaching Masters   117

◆ **Skill C** Solving equations of the form $(x + a)^2 = k$, where $k \geq 0$

Recall An equation, such as $(x + 3)^2 = 25$, is solved by using the generalization below:
If $x^2 = k$, and $k \geq 0$, then
- $x = \pm\sqrt{k}$ and
- the solutions are $\sqrt{k}$ and $-\sqrt{k}$.

When the generalization is applied to the equation $(x + 3)^2 = 25$, the result is $x + 3 = \pm\sqrt{25} = \pm 5$. Thus, $x + 3 = 5$ and $x + 3 = -5$. The solutions are 2 and -8.

◆ **Example**
Solve the equation $(x - 2)^2 - 16 = 0$.

◆ **Solution**
$(x - 2)^2 - 16 = 0$
$(x - 2)^2 = 16$
$x - 2 = \pm 4$
$x = 2 \pm 4$
The solutions are $2 + 4$, or 6, and $2 - 4$, or -2.

Solve each equation in the space provided. Round to the nearest hundredth when necessary.

10. $(x + 1)^2 = 36$ _____

11. $(x - 4)^2 = 100$ _____

12. $(x + 3)^2 = 64$ _____

13. $(x + 7)^2 - 81 = 0$ _____

14. $(x - 2)^2 - 35 = 0$ _____

15. $(x - 1)^2 - 125 = 0$ _____

Reteaching
10.3 Completing the Square

◆ **Skill A** Completing the square

Recall The area of the shaded region is represented by $x^2 + 8x$. To complete the total area of the square, the area of the non-shaded region must be added to $x^2 + 8x$. The area to be added can be found by adding the square of $\frac{1}{2}$ of the coefficient of the x-term. Since the coefficient of x is 8, $\left(\frac{1}{2} \cdot 8\right)^2$ is 4^2, or 16. Thus, the perfect square $x^2 + 8x + 16$ represents the total area of square.

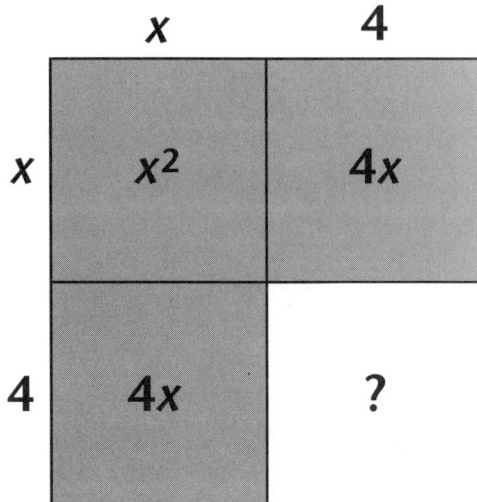

◆ **Example 1**
Complete the square for $x^2 - 4x$.

◆ **Solution**
The coefficient of x is 4. $\left(\frac{1}{2} \cdot 4\right)^2 = 2^2$, or 4
Thus, 4 completes the square.
The perfect square is $x^2 - 4x + 4$.

◆ **Example 2**
Find the minimum value for the function $f(x) = x^2 - 8x$.

◆ **Solution**
If a quadratic function is written in the form $y = (x - h)^2 + k$, the minimum value of the function is k. To complete the square for $x^2 - 8x$, 16 must be added to the expression. Add and subtract **16**. The 16 is subtracted in order to balance the equation. The result is as follows:
$y = (x^2 - 8x + \mathbf{16}) - \mathbf{16}$
$= (x - 4)^2 - \mathbf{16}$
The minimum value of f is -16.

Complete the square and find the minimum value of each function.

1. $y = x^2 + 12x$

2. $y = x^2 + 5x$

3. $y = x^2 - 20x$

4. $y = x^2 - 2x$

5. $y = x^2 - 10x$

6. $y = x^2 + 11x$

Algebra 1 — Reteaching Masters

◆ **Skill B** Finding the coordinates of the vertex of a parabola

Recall If a quadratic function is written in the form $y = (x - h)^2 + k$, the vertex of the parabola is $V(h, k)$. To find the vertex of the graph of $y = x^2 - 2x - 3$, complete the following steps:

- Group the x^2-and the x-term together. $y = (x^2 - 2x) - 3$
- Complete the square. $\left(\frac{1}{2} \cdot 2\right)^2 = 1$
- Add 1 and subtract 1. $y = (x^2 - 2x + 1) - 3 - 1$
- Write -4 in the form $y = (x - h)^2 + k$. $y = (x - 1)^2 - 4$

The vertex of the parabola is $(1, 4)$.

◆ **Example**

Rewrite the equation $y = x^2 + 3x - 2$ in the form $y = (x - h)^2 + k$. The vertex is (h, k).

◆ **Solution**

$y = x^2 + 3x - 2$ In $y = x^2 + 3x - 2$, the coefficient of x is 3.

$= \left(x^2 + 3x + \frac{9}{4}\right) - 2 - \frac{9}{4}$ $\left(\frac{1}{2} \cdot 3\right)^2 = \left(\frac{3}{2}\right)^2 = \frac{9}{4}$

$= \left(x + \frac{3}{2}\right)^2 - \frac{17}{4}$

The vertex is $\left(-\frac{3}{2}, -\frac{17}{4}\right)$.

Rewrite each function in the form $y = (x - h)^2 + k$. Find the vertex of each.

7. $y = x^2 + 2x + 1$

8. $y = x^2 - 8x + 3$

9. $y = x^2 + 4x - 3$

10. $y = x^2 - 2x + 4$

11. $y = x^2 - 12x - 36$

12. $y = x^2 + 2x - 4$

13. $y = x^2 - 3x + 6$

14. $y = x^2 - 5x - 25$

15. $y = x^2 + x + 1$

NAME _____ CLASS _____ DATE _____

Reteaching
10.4 Solving Equations of the Form $x^2 + bx + c = 0$

◆ **Skill A** Finding the zeros of a function by completing the square

Recall The zeros of the function $y = x^2 + bx + c$ are the values for which y is 0. One way to find the zeros of a function, such as $y = x^2 - 4x - 12$, is to set the expression equal to 0 and then to solve the equation by completing the square.

$$x^2 - 4x - 12 = 0$$
$$x^2 - 4x = 12$$
$$x^2 - 4x + 4 = 12 + 4$$
$$(x - 2)^2 = 16$$
$$x - 2 = \pm 4$$
$$x = 2 + 4 \text{ or } x = 2 - 4$$
$$x = 6 \quad \text{ or } x = -2$$

The zeros are 6 and -2.

◆ **Example**
Solve the equation $x^2 + 6x - 6 = 0$ by completing the square.

◆ **Solution**
$$x^2 + 6x - 6 = 0$$
$$x^2 + 6x = 6$$
$$x^2 + 6x + 9 = 15$$
$$(x + 3)^2 = 15$$
$$x + 3 = \pm\sqrt{15}$$
$$x = -3 + \sqrt{15} \text{ or } x = -3 - \sqrt{15}$$

The solutions are $-3 + \sqrt{15}$ and $-3 - \sqrt{15}$.

Solve by completing the square.

1. $y = x^2 + x - 2$
2. $y = x^2 + 3x - 10$
3. $y = x^2 + 5x + 6$

4. $y = x^2 + 2x + 1$
5. $y = x^2 - 10x + 3$
6. $y = x^2 + 4x - 3$

Algebra 1   Reteaching Masters   **121**

NAME _____ CLASS _____ DATE _____

◆ **Skill B** Finding the zeros of a function by factoring

Recall The equation $x^2 - 4x - 12 = 0$ can also be solved by factoring the expression $x^2 - 4x - 12$ and then applying the Zero Product Property. The Zero Product Property states that if $ab = 0$, then $a = 0$ or $b = 0$.

$$x^2 - 4x - 12 = 0$$
$$(x + 2)(x - 6) = 0$$
$$x + 2 = 0 \text{ or } x - 6 = 0$$
$$x = -2 \text{ or } \quad x = 6$$

The solutions are -2 and 6.

◆ **Example 1**
Solve the equation $x^2 + 2x + 1 = 0$ by factoring.

◆ **Solution**
$$x^2 + 2x + 1 = 0$$
$$(x + 1)(x + 1) = 0$$
$$x + 1 = 0 \text{ or } x + 1 = 0$$
The only solution is -1.

◆ **Example 2**
Let $y = x^2 - 4x + 3$. Find the value of x when y is 7.

◆ **Solution**
Solve $x^2 - 4x + 3 = 7$ by factoring or by completing the square.
$$x^2 - 4x + 3 = 7$$
$$x^2 - 4x = 4$$
$$x^2 - 4x + 4 = 4 + 4$$
$$(x - 2)^2 = 8$$
$$x - 2 = \pm\sqrt{8}$$
The solutions are $2 + \sqrt{8}$ and $2 - \sqrt{8}$.

Solve each equation by factoring.

7. $y = x^2 - x - 20$

8. $y = x^2 - 7x + 12$

9. $y = x^2 + 4x - 5$

Let $y = x^2 + 6x - 2$. Find the value of x for each value of y.

10. $y = -10$

11. $y = -2$

12. $y = 0$

Reteaching Masters   **Algebra 1**

NAME _____ CLASS _____ DATE _____

Reteaching
10.5 The Quadratic Formula

◆ **Skill A** Using the quadratic formula to solve equations

Recall The solutions for a quadratic equation written in the form $ax^2 + bx + c = 0$, where $a \neq 0$, can be found by using the quadratic formula, $x = \dfrac{-b \pm \sqrt{b^2 - 4ac}}{2a}$

◆ **Example**
Use the quadratic formula to solve $x^2 - 8x + 15 = 0$ for x.

◆ **Solution**
For $x^2 - 8x + 15 = 0$, a is 1; b is -8, and c is 15. Substitute these values in the quadratic formula.

$$x = \dfrac{-(-8) \pm \sqrt{(-8)^2 - (4)(1)(15)}}{(2)(1)}$$
$$= \dfrac{8 \pm \sqrt{64 - 60}}{2}$$
$$= \dfrac{8 \pm \sqrt{4}}{2}$$
$$= \dfrac{8 \pm 2}{2}$$
$x = 3$ or $x = 5$
The solutions are 3 and 5.

Use the quadratic formula to solve each equation.

1. $x^2 - 5x + 4 = 0$ 2. $x^2 - 2x - 24 = 0$ 3. $x^2 + 6x + 9 = 0$

4. $x^2 + 3x - 10 = 0$ 5. $2x^2 - x - 6 = 0$ 6. $2x^2 + x - 4 = 0$

Algebra 1   Reteaching Masters   **123**

NAME _____ CLASS _____ DATE _____

◆ **Skill B** Using the quadratic formula to find the zeros of quadratic functions

Recall The zeros of a quadratic function written in the form $y = ax^2 + bx + c = 0$, where $a \neq 0$, can be found by using the quadratic formula.

◆ **Example**
Use the quadratic formula to find the zeros of $y = 2x^2 - 5x - 3$.

◆ **Solution**
For $2x^2 - 5x - 1$, a is 2; b is -5, and c is -3. Substitute these values in the quadratic formula.
$$\frac{-(-5) \pm \sqrt{(-5)^2 - (4)(2)(-3)}}{(2)(2)} = \frac{5 \pm \sqrt{25 + 24}}{4} = \frac{5 \pm \sqrt{49}}{4} = \frac{5 \pm 7}{4}$$
The zeros are $-\frac{1}{2}$ and 3.

Use the quadratic formula to find the zeros of each function.

7. $y = x^2 + 2x - 8$
8. $y = 2x^2 - x - 15$
9. $y = 4x^2 - 8x + 3$

◆ **Skill C** Using the discriminant to determine the number of solutions

Recall When a quadratic equation is written in the form $ax^2 + bx + c = 0$, where $a \neq 0$, the expression $b^2 - 4ac$ is called the *discriminant* of the quadratic formula.
If $b^2 - 4ac > 0$, there are two solutions.
If $b^2 - 4ac = 0$, there is one solution.
If $b^2 - 4ac < 0$, there are no real number solutions.

◆ **Example**
What does the discriminant tell you about $3x^2 - 2x + 9 = 0$?

◆ **Solution**
For $3x^2 - 2x + 9 = 0$, a is 3; b is -2, and c is 9.
Thus, $b^2 - 4ac = (-2)^2 - (4)(3)(9) = 4 - 108 = -104$
$-104 < 0$, so the equation $3x^2 - 2x + 9 = 0$ has no real solutions.

Give the value of each discriminant. What does the discrimant tell you about the function?

10. $y = 4x^2 + 4x + 1$
11. $y = x^2 + 5x + 4$
12. $y = x^2 + 5x + 8$

Reteaching

10.6 Graphing Quadratic Inequalities

◆ **Skill A** Solving quadratic inequalities

Recall The solution to a quadratic inequality, such as $x^2 + 3x - 10 \leq 0$, can be graphed on a number line. The first step is to find the solution to $x^2 + 3x - 10 = 0$ by factoring, by completing the square, or by using the quadratic formula. Because $x^2 + 3x - 10 = (x + 5)(x - 2)$, the solutions to $x^2 + 3x - 10 = 0$ are -5 and 2. The points representing -5 and 2 divide the number line into three regions.

```
Region 1 | Region 2 | Region 3
────────┼──────────┼────────→ x
        -5    0    2
```

The region between -5 and 2 represents the solutions that make the expression $x^2 + 3x - 10$ less than 0. The regions to the left of -5 and to the right of 2 represent the solutions that make the expression greater than 0. Thus, the solution for the inequality is a closed line segment between -5 and 2. The segment is closed because the inequality contains the equal sign.

```
←──●━━━━━━━━●──→ x
   -5    0    2
```

◆ **Example**
Solve the quadratic inequality $x^2 + 5x - 14 > 0$. Graph the solution on a number line.

◆ **Solution**
First solve the equality $x^2 + 5x - 14 = 0$.
$x^2 + 5x - 14 = 0$
$(x + 7)(x - 2) = 0$
The solutions to the equality are -7 and 2. The regions to the left of -7 and to the right of 2 are the solutions to $x^2 + 5x - 14 > 0$. Since the equal sign is not part of the inequality, the graph will be two open rays.

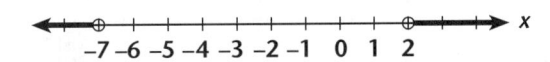

```
←─○─┼─┼─┼─┼─┼─┼─┼─┼─○─┼─→ x
  -7 -6 -5 -4 -3 -2 -1 0  1  2
```

Solve each quadratic inequality by using the Zero Product Property. Graph each solution on a number line.

1. $x^2 + 3x + 2 < 0$ _____

   ←─┼─┼─┼─┼─┼─┼─┼─┼─┼─┼─┼─→ x

2. $x^2 - 8x + 15 \geq 0$ _____

   ←─┼─┼─┼─┼─┼─┼─┼─┼─┼─┼─┼─→ x

3. $x^2 - 9 > 0$ _____

   ←─┼─┼─┼─┼─┼─┼─┼─┼─┼─┼─┼─→ x

4. $x^2 + x - 6 \leq 0$ _____

   ←─┼─┼─┼─┼─┼─┼─┼─┼─┼─┼─┼─→ x

Algebra 1   Reteaching Masters

◆ Skill B Graphing quadratic inequalities

Recall To graph a quadratic inequality, such as $y \leq x^2 + 2x - 8$, first graph the equality $y = x^2 + 2x - 8$. The solution to an inequality containing $<$ consists of all the points outside the parabola. The solution to an inequality containing $>$ consists of all points inside the parabola. If the inequality contains $\leq$ or $\geq$, the parabola is drawn with a solid line. Otherwise, the parabola is drawn with a dashed line. The graph of $y \leq x^2 + 2x - 8$ is shown.

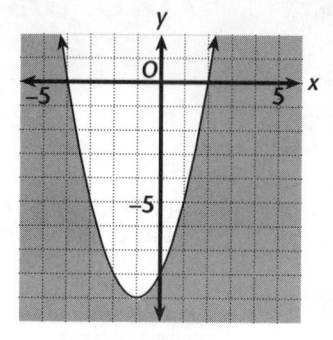

◆ **Example**
Graph the quadratic inequality $y \geq x^2 - 2x + 1$. Shade the solution region.

◆ **Solution**
First graph the quadratic inequality. Draw the parabola with a solid line. To be sure which region should be shaded, test the point $(0, 0)$ in the inequality. Because $0 \geq 1$ is false, shade the region that does not contain the point $(0, 0)$.

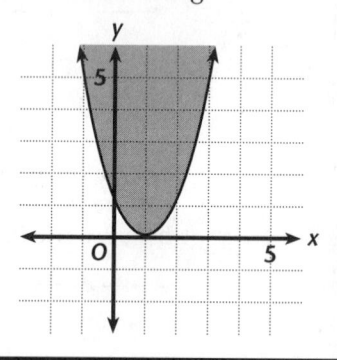

Graph each quadratic inequality. Shade the solution region.

5. $y > x^2 - 6x$

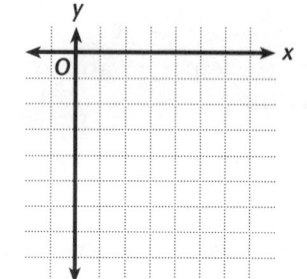

6. $y < x^2 + 4x + 4$

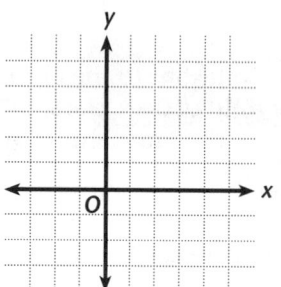

7. $y \geq x^2 - 4x + 5$

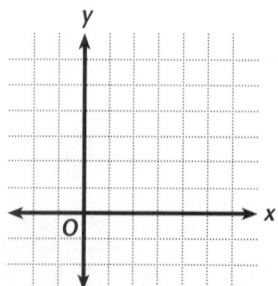

8. $y \leq x^2 - 3x + 2$

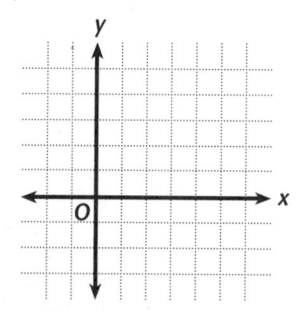

Reteaching
11.1 Inverse Variation

◆ **Skill A** Solving inverse-variation equations

Recall If $y = \frac{k}{x}$ or $xy = k$ (where $k \neq 0$ and $x \neq 0$), then y varies inversely as x; that is, as x increases, y decreases, and as x decreases, y increases. The constant of variation is the value of k. The graph of an inverse-variation equation is a hyperbola.

◆ **Example 1**
Determine whether the equation $y = 2x$ describes inverse variation.

◆ **Solution**
In the equation $y = 2x$, as x increases, y also increases, and its graph is a straight line. Therefore, the equation $y = 2x$ does not describe inverse variation.

◆ **Example 2**
Suppose that y varies inversely as x. If y is 10 when x is 6, find y when x is 4.

◆ **Solution**
Because y varies inversely as x, $y = \frac{k}{x}$.

$k = (6)(10)$, or 60
The constant of variation is 60, so $4y$ must also be 60.
If $4y = 60$, y must equal 15.

Determine whether each equation describes an inverse variation.

1. $dt = 150$

2. $p = \frac{-10}{q}$

3. $2a - 5 = b$

Suppose that y varies inversely as x. For each problem, if y is

4. 6 when x is 12, find x when y is 9.

5. 25 when x is 4, find x when y is 10.

6. $\frac{2}{3}$ when x is 15, find x when y is 2.

Algebra 1 — Reteaching Masters 127

NAME _____ CLASS _____ DATE _____

◆ **Skill B** Solving inverse-variation problems

Recall If an equation involves inverse variation, find the constant of variation and then find the unknown quantity.

◆ **Example**
Time traveled varies inversely as the rate of travel. You can drive 8 hours at 50 miles per hour. How many hours will it take to make the same trip at 40 miles per hour?

◆ **Solution**
Distance is the constant of variation. Since distance is equal to the product of rate and time, $d = rt$ is the inverse-variation equation to be solved.
$d = (8)(50) = 400$
The constant of variation is 400.
Thus, $40t = 400$ and $t = 10$.
It will take 10 hours to travel 400 miles at 40 miles per hour.

Solve each problem.

7. The length of the base of a triangle with a constant area varies inversely as the triangle's height. When the base is 36 inches, the height is 8 inches. Find the length of the base when the height is 12 inches.

8. The current in an electric circuit varies inversely as the resistance. When the current is 60 amps, the resistance is 12 ohms. Find the current when the resistance is 16 ohms.

9. The number of days to finish a job varies inversely as the number of people completing the job. If it takes 8 people 16 days to finish the job, how long will it take 12 people to finish the same job?

10. The length of a rectangle with a constant area varies inversely as the rectangle's width. When the length is 10 meters, the width is 5 meters. Find the length when the width is 2 meters.

128   Reteaching Masters   Algebra 1

Reteaching
11.2 Rational Expressions and Functions

◆ **Skill A** Evaluating rational expressions

Recall A rational expression is an expression of the form $\frac{P}{Q}$, where P and Q are polynomials and $Q \neq 0$.

◆ **Example 1**
Identify the values for which the rational expression $\frac{x+3}{x^2-4x+4}$ is undefined.

◆ **Solution**
The rational expression is undefined when $x^2 - 4x + 4 = 0$.
Because $x^2 - 4x + 4 = 0$ when $(x-2)^2 = 0$, the expression $x^2 - 4x + 4$ equals 0 when x is 2.
Thus, $\frac{x+3}{x^2-4x+4}$ is undefined when $x = 2$.

◆ **Example 2**
Evaluate the rational expression $\frac{x-1}{x^2-4}$ for $x = 3$ and $x = 2$.

◆ **Solution**
For $x = 3$: $\frac{x-1}{x^2-4} = \frac{3-1}{9-4} = \frac{2}{5}$

For $x = 2$: $\frac{x-1}{x^2-4} = \frac{2-1}{4-4} = \frac{1}{0}$ Thus, $\frac{x-1}{x^2-4}$ is undefined when $x = 2$.

For what values of the variable is each rational expression undefined?

1. $\frac{3x+4}{x+1}$

2. $\frac{5x}{x^2+5x-6}$

3. $\frac{x^2+3x+3}{x^2-9}$

Evaluate each rational expression for $x = 3$ and $x = -1$. Write *undefined* if appropriate.

4. $\frac{5x+2}{3x-12}$

5. $\frac{x+3}{x^2+6x+9}$

6. $\frac{x^2-x-2}{x^2+4x+3}$

Algebra 1 Reteaching Masters **129**

◆ **Skill B** Transforming rational expressions

Recall The parent function for a rational function is $f(x) = \frac{1}{x}$, where $x \neq 0$.

◆ **Example**
Describe the transformations applied to the parent function $f(x) = \frac{1}{x}$ in order to produce the graph of $g(x) = \frac{1}{x} + 1$.

◆ **Solution**
Graph $f(x)$ and $g(x)$.

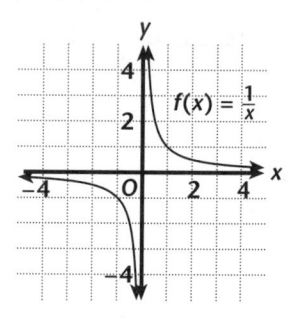

$f(x) = \frac{1}{x}$ 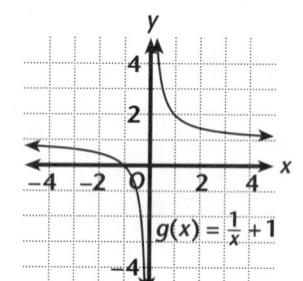 $g(x) = \frac{1}{x} + 1$

The graph of $f(x)$ has been shifted 1 unit up in order to produce the graph of $g(x)$.

Describe the transformations applied to the parent function $f(x) = \frac{1}{x}$ in order to produce the graph of each rational function below.

7. $g(x) = \frac{1}{x} - 3$ _____

8. $g(x) = \frac{1}{x-1}$ _____

9. $g(x) = \frac{1}{x+1} + 2$ _____

10. $g(x) = \frac{2}{x+2}$ _____

11. $g(x) = \frac{3}{x+5} + 2$ _____

12. $g(x) = \frac{1}{x-3} - 5$ _____

NAME _____ CLASS _____ DATE _____

Reteaching
11.3 Simplifying Rational Expressions

◆ **Skill A** Factoring out monomials

Recall A rational expression is an expression of the form $\frac{P}{Q}$, where P and Q are polynomials and $Q \neq 0$. A rational expression is in simplest form when the numerator and denominator have no common factors other than 1 or -1. Be sure to check for undefined terms before simplifying the expression.

◆ **Example 1**
Simplify the rational expression $\frac{6c - 9}{12c}$.

◆ **Solution**
First factor the numerator and denominator.

$$\frac{6c - 9}{12c} = \frac{3(2c - 3)}{(2)(2)(3)c}$$
$$= \frac{3}{3} \cdot \frac{2c - 3}{4c}$$

Rewrite to show any fractions equal to 1.

Thus, the simplified form is $\frac{2c - 3}{4c}$, where $c \neq 0$.

◆ **Example 2**
Simplify the rational expression $\frac{5m^2}{5m + 10m^2}$.

◆ **Solution**
First factor the numerator and denominator.

$$\frac{5m^2}{5m + 10m^2} = \frac{5 \cdot m \cdot m}{5 \cdot m \cdot (1 + 2m)}$$
$$= \frac{5m}{5m} \cdot \frac{m}{2m + 1}$$

Rewrite to show any fractions equal to 1.

Thus, the simplified form is $\frac{m}{2m + 1}$, where $m \neq 0$ and $m \neq -\frac{1}{2}$.

Simplify each expression and state any restrictions on the variables.

1. $\frac{14t}{7t - 7}$

2. $\frac{3m + 9}{6m - 12}$

3. $\frac{4m^2}{8m + 12m^2}$

4. $\frac{3y + 6}{y^2 + 4y + 2}$

5. $\frac{3x + 3}{x^2 + 2x + 1}$

6. $\frac{4r - 12}{r^2 - 6r + 9}$

Algebra 1   Reteaching Masters   **131**

◆ Skill B Factoring out binomials

Recall A rational expression is in simplest form when the numerator and denominator have no common factors other than 1 or -1.

◆ **Example**
Simplify the rational expression $\dfrac{x-1}{x^2 - 3x + 2}$.

◆ **Solution**
First factor the numerator and denominator.
$$\dfrac{x-1}{x^2 - 3x + 2} = \dfrac{x-1}{(x-1)(x-2)}$$

Write the restrictions on the variable. $x \neq 1$ and $x \neq 2$

Rewrite to show any fractions equal to 1.
$$= \dfrac{x-1}{x-1} \cdot \dfrac{1}{x-2}$$

Thus, the simplified form is $\dfrac{1}{x-2}$, where $x \neq 1$ and $x \neq 2$.

Simplify each expression and state any restrictions on the variables.

7. $\dfrac{m^2 + 3m + 2}{m + 1}$

8. $\dfrac{a - 2}{a^2 - 5a + 6}$

9. $\dfrac{3b + 6}{b^2 + 4b + 4}$

10. $\dfrac{r^2 - r}{r^2 - 1}$

11. $\dfrac{x^2 - 1}{x^2 - x}$

12. $\dfrac{x^2 + x - 6}{x^2 - 6x + 8}$

13. The area of a rectangular flower bed is represented by $x^2 + 3x - 10$. One side of the flower bed is represented by $x + 5$. What is the length of the other side of the flower bed?

132   Reteaching Masters   Algebra 1

Reteaching
11.4 Operations with Rational Expressions

♦ Skill A Multiplying rational expressions

Recall To multiply two rational expressions, multiply the numerators and multiply the denominators. Then simplify the results. List any restrictions.

♦ **Example**
Find the product $\dfrac{3a+3}{a} \cdot \dfrac{a-3}{a+1}$.

♦ **Solution**

Multiply numerators and denominators. $\dfrac{3a+3}{a} \cdot \dfrac{a-3}{a+1} = \dfrac{(3a+3)(a-3)}{a(a+1)}$

Factor all expressions. $= \dfrac{3(a+1)(a-3)}{a(a+1)}$

Rewrite to show any fractions equal to 1. $= \dfrac{a+1}{a+1} \cdot \dfrac{3(a-3)}{a}$

Simplify and note any restrictions. $= \dfrac{3(a-3)}{a}$, where $a \neq 0$ and $a \neq -1$

Find each product.

1. $\dfrac{2t-2}{t} \cdot \dfrac{t^2}{t-1}$

2. $\dfrac{12a-18}{18a} \cdot \dfrac{2a}{2a-3}$

3. $\dfrac{5}{c+3} \cdot \dfrac{c^2+6c+9}{10}$

4. $\dfrac{20d}{d^2+8d+15} \cdot \dfrac{d+3}{5d}$

5. $\dfrac{x^2-9}{4} \cdot \dfrac{8}{x+3}$

6. $\dfrac{y^2-y-2}{y+2} \cdot \dfrac{y^2+y-2}{y-2}$

Algebra 1   Reteaching Masters   **133**

NAME _____ CLASS _____ DATE _____

◆ **Skill B** Adding and subtracting rational expressions

Recall Rewrite all rational expressions with common denominators. Add or subtract the numerators. Keep the same denominator. Then simplify the results. List any restrictions.

◆ **Example**
Find the difference $\dfrac{b}{b+3} - \dfrac{4}{b+2}$.

◆ **Solution**
Multiply numerators and denominators by fractions that are equal to 1 and that will result in common denominators.
$$\dfrac{b+2}{b+2} \cdot \dfrac{b}{b+3} - \dfrac{b+3}{b+3} \cdot \dfrac{4}{b+2}$$

Multiply the rational expressions.
$$\dfrac{b^2+2b}{(b+2)(b+3)} - \dfrac{4b+12}{(b+2)(b+3)}$$

Subtract the numerators and simplify.
$$\dfrac{b^2-2b-12}{(b+2)(b-3)}$$

Find each sum or difference.

7. $\dfrac{x}{4} + \dfrac{2x}{5}$

8. $\dfrac{7}{2b} + \dfrac{8}{3b}$

9. $\dfrac{t}{4t^2} - \dfrac{5}{6t}$

10. $\dfrac{2m}{m-1} + \dfrac{m}{2m-2}$

11. $\dfrac{3c}{3c-12} - \dfrac{c}{2c-8}$

12. $\dfrac{y}{y+2} + \dfrac{2y}{y-2}$

NAME _____ CLASS _____ DATE _____

Reteaching
11.5 Solving Rational Equations

◆ **Skill A** Using the common denominator method to solve rational equations

Recall To solve an equation containing rational expressions, first multiply each side of the equality by the common denominator of all the rational expressions. This will clear the equations of fractions. Solve the resulting equation. Check the solutions and the restrictions.

◆ **Example**

Solve. **a.** $\frac{x}{3} + \frac{x+1}{4} = \frac{17}{12}$ **b.** $\frac{3}{x-1} - 2 = -1$

◆ **Solution**

a. Multiply each side of the equation by 12.

$$12\left(\frac{x}{3} + \frac{x+1}{4}\right) = 12 \cdot \frac{17}{12}$$

$$4x + 3x + 3 = 17$$
$$7x = 14$$
$$x = 2$$

b. Multiply each side of the equation by $x - 1$.

$$(x-1)\left(\frac{3}{x-1} - 2\right) = (x-1)(-1), \text{ where } x \neq 1$$

$$(x-1)\left(\frac{3}{x-1}\right) - (x-1)(2) = (x-1)(-1)$$
$$3 - 2x + 2 = -x - 1$$
$$x = 4$$

Solve each rational equation by using the lowest common denominator and state any restrictions on the variable.

1. $\frac{a}{3} - \frac{a}{2} = 1$

2. $\frac{5}{d} + \frac{3}{d^2} = \frac{13}{4}$

3. $\frac{h-2}{h} - \frac{h-3}{h-6} = \frac{1}{h}$

4. $\frac{7}{x-4} - \frac{5}{x-2} = 0$

Algebra 1   Reteaching Masters   135

◆ Skill B Solving work problems by using rational equations

Recall Work problems deal with situations in which people work at different rates in order to complete a job. The formula usually used for work problems is
rate of work · time = work done, or $r \cdot t = w$.

For example, suppose that it takes 5 hours to paint one room. Then $\frac{1}{5}$ of the job can be completed in 1 hour; $2 \cdot \frac{1}{5}$, or $\frac{2}{5}$, of the job can be completed in 2 hours, and $n \cdot \frac{1}{5}$, or $\frac{n}{5}$, of the job can be completed in n hours. Completing the whole job is $5 \cdot \frac{1}{5}$, or 1 job.

◆ Example
Paul can clean a room in 60 minutes. Pat can clean the same room in 30 minutes. If Paul and Pat work together, how long will it take to clean the room?

◆ Solution
Let t represent the number of minutes that it will take working together.

Paul's rate of work is $\frac{1}{60}$ of the job per minute. His part is shown as $\frac{t}{60}$.

Pat's rate of work is $\frac{1}{30}$ of the job per minute. Her part is shown as $\frac{t}{30}$.

The total of the two parts of the job equals the whole job, or 1.

Thus, $\frac{t}{60} + \frac{t}{30} = 1$

$60\left(\frac{t}{60} + \frac{t}{30}\right) = 60 \cdot 1$

$t + 2t = 60$

$3t = 60$

$t = 20$

Together, Paul and Pat can clean the room in 20 minutes.

Solve each problem.

5. It takes 4 hours to fill a swimming pool with water from one hose. It takes 6 hours with a smaller hose. How long will it take to fill the pool if both hoses are used at the same time?

6. Working together, Sergi and Anita can wash all of the windows in their house in 3 hours. Working by himself, Sergi can wash the windows in 5 hours. How long will it take Anita to wash the windows if she does the job alone?

Reteaching
11.6 Proof in Algebra

◆ **Skill A** Using proofs to solve equations

Recall A statement written in if-then form is called a conditional. The *if* clause is assumed to be true, and the properties of algebra are used to show that the *then* clause follows logically.

◆ **Example**
Use the properties of algebra to prove the following conditional:
If $\frac{10}{x+1} = \frac{15}{12}$, then $x = 7$.

◆ **Solution**
Start with the *if* clause and give a reason for each step until you reach the *then* clause.

$\frac{10}{x+1} = \frac{15}{12}$	Given
$15(x + 1) = 12(10)$	Cross products
$15x + 15 = 120$	Distributive Property
$15x = 105$	Subtraction Property
$x = 7$	Division Property

Write a proof for each conjecture. Give a reason for each step. Let all variables represent real numbers.

1. $3x - 6 = 15$

2. $8x^2 = 72$

3. $x^2 - x = 12$

4. $\frac{5}{x} = \frac{x-3}{2}$

Algebra 1   Reteaching Masters

◆ Skill B Proving theorems

Recall When a statement can be proven for every case, the conditional is called a theorem.

◆ Example
Use the properties of algebra to prove the following theorem:
   The square of an odd integer is always an odd integer.

◆ Solution
Let x represent any integer.	Given
$2x$ is an even integer.	Any integer in the form $2x$ is even.
$2x + 1$ is odd.	Any odd integer is an even number plus 1.
$(2x + 1)^2 = 2x^2 + 4x + 1$	Squaring a binomial
$\qquad\quad = 2(x^2 + 2x) + 1$	Distributive Property

The integer represented by $2(x^2 + 2x) + 1$ is odd because it is an even number, $2(x^2 + 2x)$, plus 1. Thus, any odd integer squared will always result in an odd integer.

Prove each theorem. Give a reason for each step.

5. The sum of two odd numbers is even.

6. The product of consecutive odd and even numbers is even.

7. The sum of consecutive integers is odd.

8. The sum of two multiples of 3 is also a multiple of 3.

NAME _____ CLASS _____ DATE _____

Reteaching
12.1 Operations With Radicals

◆ **Skill A** Simplifying radicals

Recall Radical expressions are in simplest radical form if the expression under the radical sign contains no perfect squares and there are no radicals in the denominator.

◆ **Example**
Simplify each of the following:

a. $\sqrt{18}$ b. $\dfrac{\sqrt{4c^3}}{\sqrt{50c^2}}$

◆ **Solution**

a. $\sqrt{18} = \sqrt{9 \cdot 2} = \sqrt{3^2 \cdot 2} = \sqrt{3^2} \cdot \sqrt{2} = 3\sqrt{2}$

b. $\dfrac{\sqrt{4c^3}}{\sqrt{50c^2}} = \dfrac{\sqrt{2^2 \cdot c^2 \cdot c}}{\sqrt{5^2 \cdot 2 \cdot c^2}} = \dfrac{2c\sqrt{c}}{5c\sqrt{2}} = \dfrac{2\sqrt{c}}{5\sqrt{2}}$

To further simplify the expression, multiply the denominator by a radical that will make a perfect square.

$\dfrac{2\sqrt{c}}{5\sqrt{2}} = \dfrac{2\sqrt{c}}{5\sqrt{2}} \cdot \dfrac{\sqrt{2}}{\sqrt{2}} = \dfrac{2\sqrt{2}\sqrt{c}}{5 \cdot \sqrt{4}} = \dfrac{2\sqrt{2}\sqrt{c}}{5 \cdot 2} = \dfrac{\sqrt{2}\sqrt{c}}{5} = \dfrac{\sqrt{2c}}{5}$

Simplify each radical expression.

1. $\sqrt{48}$ _____   2. $\sqrt{200b^2}$ _____

3. $\dfrac{\sqrt{72c^2}}{\sqrt{36c^3}}$ _____   4. $\dfrac{\sqrt{12a^4}}{\sqrt{20a^2}}$ _____

◆ **Skill B** Adding radicals

Recall Radical expressions with like radicands can be added by using the Distributive Property.

◆ **Example**
Simplify $\sqrt{8} + \sqrt{18}$.

◆ **Solution**
$\sqrt{8} + \sqrt{18} = \sqrt{2 \cdot 4} + \sqrt{2 \cdot 9} = 2\sqrt{2} + 3\sqrt{2} = (2 + 3)\sqrt{2} = 5\sqrt{2}$

Simplify.

5. $2\sqrt{6} + 5\sqrt{6}$ _____   6. $3\sqrt{3} + \sqrt{27}$ _____

7. $\sqrt{50} + \sqrt{98}$ _____   8. $3\sqrt{2} + \sqrt{8} + \sqrt{2}$ _____

Algebra 1   Reteaching Masters   **139**

NAME _____ CLASS _____ DATE _____

◆ Skill C Subtracting radicals

Recall Radical expressions with like radicands can also be subtracted.

◆ **Example**
Simplify $3\sqrt{2} - \sqrt{2} - \sqrt{12}$.

◆ **Solution**
$$3\sqrt{2} - \sqrt{2} - \sqrt{12} = (3\sqrt{2} - \sqrt{2}) - 2\sqrt{3}$$
$$= ((3-1)\sqrt{2}) - 2\sqrt{3}$$
$$= 2\sqrt{2} - 2\sqrt{3}$$

Simplify each radical expression.

9. $3\sqrt{5} - 7\sqrt{5}$ _____

10. $5\sqrt{8} - \sqrt{72}$ _____

11. $8\sqrt{75} - 9\sqrt{3}$ _____

12. $\sqrt{6} - \sqrt{24} - \sqrt{36}$ _____

◆ Skill D Multiplying radicals

Recall Expressions containing radicals can be multiplied by using the Distributive Property or the FOIL method.

◆ **Example**
Simplify $(2 + \sqrt{3})(5 - \sqrt{3})$.

◆ **Solution**
$$(2 + \sqrt{3})(5 - \sqrt{3}) = 2 \cdot 5 - 2 \cdot \sqrt{3} + \sqrt{3} \cdot 5 - \sqrt{3} \cdot \sqrt{3}$$
$$= 10 - 2\sqrt{3} + 5\sqrt{3} - 3$$
$$= 7 + 3\sqrt{3}$$

Simplify.

13. $\sqrt{2}(\sqrt{6} + 5)$

14. $\sqrt{3}(5 - \sqrt{12})$

15. $\sqrt{6}(\sqrt{3} + \sqrt{2})$

16. $(2 + \sqrt{5})(2 - \sqrt{5})$

17. $(4 + \sqrt{12})(3 - \sqrt{2})$

18. $(\sqrt{3} + \sqrt{5})^2$

NAME _____ CLASS _____ DATE _____

Reteaching

12.2 Square-Root Functions and Radical Equations

◆ **Skill A** Solving equations that contain radicals

Recall To solve an equation such as $\sqrt{x + 3} = 2$, square each side of the equation to eliminate the radical sign. Then solve and check.

◆ **Example 1**
Solve $\sqrt{x + 3} = 2$.

◆ **Solution**
$(\sqrt{x + 3})^2 = 2^2$
$x + 3 = 4$
$x = 1$
The solution is 1 because $\sqrt{1 + 3} = 2$.

◆ **Example 2**
Solve $\sqrt{2x + 8} = x$.

◆ **Solution**
$(\sqrt{2x + 8})^2 = x^2$
$2x + 8 = x^2$
$x^2 - 2x - 8 = 0$
$(x - 4)(x + 2) = 0$
$x = 4$ or $x = -2$

Substitute 4 and -2 in the original equation to check the solutions.
$\sqrt{2(4) + 8} = \sqrt{8 + 8} = \sqrt{16} = 4$, so 4 is a solution.
$\sqrt{2(-2) + 8} = \sqrt{-4 + 8} = \sqrt{4} \neq -2$, so -2 is not a solution.
Thus, 4 is the only real solution of $\sqrt{2x + 8} = x$.

Solve each equation algebraically. Be sure to check your solution.

1. $\sqrt{x - 1} = 3$

2. $\sqrt{2x + 3} = 5$

3. $\sqrt{x - 7} = 4$

4. $\sqrt{x + 8} = 4$

5. $\sqrt{x + 20} = x$

6. $\sqrt{4x - 4} = x$

Algebra 1

Reteaching Masters **141**

NAME _____ CLASS _____ DATE _____

◆ Skill B Using the square-root generalization

Recall An equation such as $x^2 + 6x + 9 = 49$ can be solved by using the square-root generalization. If $x^2 = k$, $x = \sqrt{k}$ or $x = -\sqrt{k}$. Because the left side of the equation is a perfect square, the equation can be written in the form $(x + 3)^2 = 49$.

$(x + 3)^2 = 49$
$x + 3 = \pm\sqrt{49}$
$x + 3 = 7$ or $x + 3 = -7$
$x = 4$ or $x = -10$

The solutions are 4 and -10 because $(4 + 3)^2 = 49$ and $(-10 + 3)^2 = 49$.

◆ **Example**
Solve $x^2 - 2x + 1 = 9$.

◆ **Solution**
$x^2 - 2x + 1 = 9$
$(x - 1)^2 = 9$
$x - 1 = \pm\sqrt{9}$
$x - 1 = 3$ or $x - 1 = -3$
$x = 4$ or $x = -2$
The solutions are 4 and -2, which check in the original equation.

Solve each equation and write the solutions in simplest form.

7. $x^2 = 240$

8. $\sqrt{x} + 3 = 8$

9. $\sqrt{5x - 1} = 3$

10. $\sqrt{6 - x} = 2$

11. $x^2 + 2x + 1 = 16$

12. $(x + 3)^2 = 20$

13. $\sqrt{2x + 1} = 5$

14. $\sqrt{2x} = x$

15. $\sqrt{x^2 - 3x + 6} = x + 1$

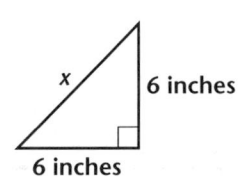

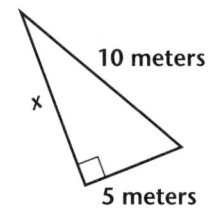

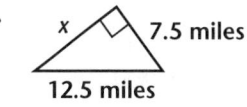

Reteaching
12.3 The Pythagorean Theorem

◆ **Skill A** Using the Pythagorean Theorem

Recall In a right triangle, the square of the hypotenuse is equal to the sum of the squares of the legs.

$$c^2 = a^2 + b^2$$
$$a^2 = c^2 - b^2$$
$$b^2 = c^2 - a^2$$

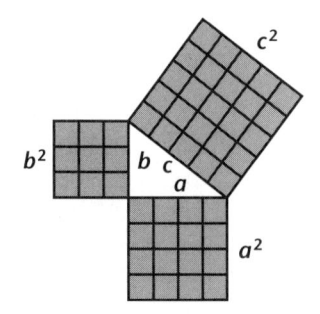

◆ **Example**
Find the hypotenuse of the right triangle.

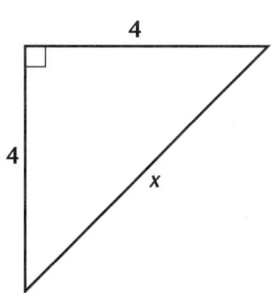

◆ **Solution**
Let x equal the length of the hypotenuse.
$x^2 = 4^2 + 4^2$
$x^2 = 32$
$x = \pm\sqrt{32}$
$x = \pm\sqrt{16 \cdot 2}$
$x = \pm 4\sqrt{2}$

Since the hypotenuse must be positive, the hypotenuse is $4\sqrt{2}$, or about 5.66.

Find the unknown length in each right triangle.

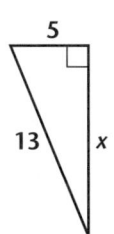

1. (triangle with legs 3 and 4, hypotenuse x)

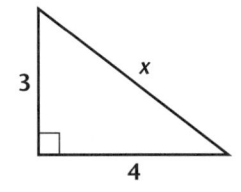

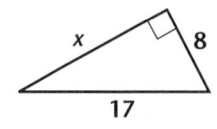

2. (right triangle with leg 5, hypotenuse 13, other leg x)

3. (right triangle with leg 8, hypotenuse 17, other leg x)

4. (right triangle with legs 6 inches and 6 inches, hypotenuse x)

5. (right triangle with legs 5 meters and x, hypotenuse 10 meters)

6. (right triangle with leg 7.5 miles, hypotenuse 12.5 miles, other leg x)

Algebra 1 Reteaching Masters **143**

◆ **Skill B** Solving right-triangle problems

Recall The Pythagorean Theorem can be used to solve many problems involving right triangles.

◆ **Example**
The top of a 12-foot ladder reaches the top of a wall that is 10 feet tall. How far from the wall is the ladder?

◆ **Solution**
Draw a diagram. Note that the wall and the ground form a right angle. To find the distance, x, use the Pythagorean Theorem.
$$a^2 = c^2 - b^2$$
$$x^2 = 12^2 - 10^2$$
$$x^2 = 144 - 100$$
$$x^2 = 44$$
$$x = \pm\sqrt{44}, \text{ or } \pm 2\sqrt{11}, \text{ or } \approx \pm 6.6$$
Since the distance must be positive, the ladder is about 6.6 feet from the wall.

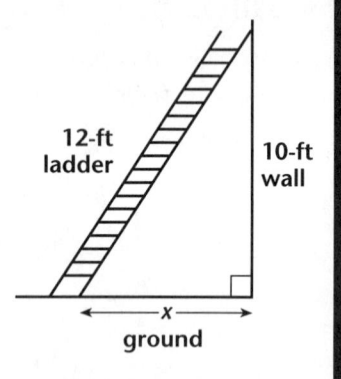

Solve each problem. Round the answer to the nearest hundredth, if necessary.

7. A ship leaves the marina and travels west for 45 miles. The ship then turns north and travels for 60 miles. How far is the ship from the marina? _____

8. A rectangular swimming pool is 25 meters long and 12 meters wide.

   What is the distance from corner to corner? _____

9. A square-shaped garden has a diagonal that measures 15 feet. What are

   the dimensions of the garden? _____

Reteaching
12.4 The Distance Formula

◆ **Skill A** Using the distance formula

Recall The distance d between two points with coordinates $P(x_1, y_1)$ and $Q(x_2, y_2)$ is $d = \sqrt{(x_2 - x_1)^2 + (y_2 - y_1)^2}$.

◆ **Example 1**
Use the distance formula to find the distance between $A(6, 9)$ and $B(-2, 5)$.

◆ **Solution**
$d = \sqrt{(-2 - 6)^2 + (5 - 9)^2}$
$= \sqrt{64 + 16}$
$= \sqrt{80}$
$= \sqrt{16 \cdot 5}$
$= 4\sqrt{5}$

The distance between the points is $4\sqrt{5}$ units.

◆ **Example 2**
Use the distance formula and the Pythagorean Theorem to prove that triangle ABC is a right triangle.

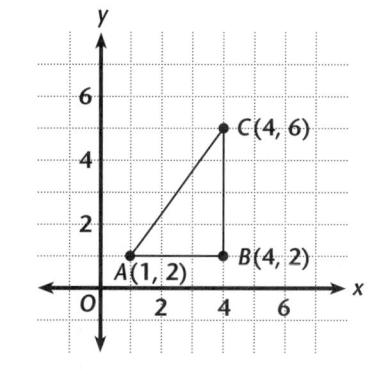

◆ **Solution**
Find the length of the hypotenuse and of each leg.
$AB = \sqrt{(4 - 1)^2 + (2 - 2)^2} = \sqrt{3^2} = 3$
$BC = \sqrt{(4 - 4)^2 + (6 - 2)^2} = \sqrt{4^2} = 4$
$AC = \sqrt{(4 - 1)^2 + (6 - 2)^2} = \sqrt{25} = 5$

$3^2 + 4^2 = 5^2$

The square of the hypotenuse is equal to the sum of the squares of the legs, so triangle ABC is a right triangle.

Find the distance between each pair of points.

1. $A(4, -3)$ and $B(-4, 3)$

2. $C(-2, -5)$ and $D(-2, 4)$

3. $E(7, 10)$ and $F(6, 2)$

4. Points $L(-2, 3)$, $M(3, -2)$, and $N(6, 1)$ form a triangle. Show whether triangle LMN is a right triangle.

NAME _____ CLASS _____ DATE _____

◆ **Skill B** Using the midpoint formula

Recall The midpoint M of a segment with endpoints $P(x_1, y_1)$ and $Q(x_2, y_2)$ has coordinates $M\left(\dfrac{x_1 + x_2}{2}, \dfrac{y_1 + y_2}{2}\right)$.

◆ **Example 1**
Find the midpoint of $\overline{ST}$.

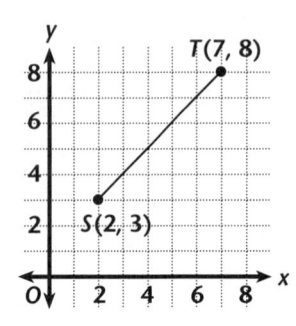

◆ **Solution**
Use the midpoint formula. Point S is $(2, 3)$ and point T is $(7, 8)$.

$$M\left(\dfrac{2 + 7}{2}, \dfrac{3 + 8}{2}\right) = M\left(\dfrac{9}{2}, \dfrac{11}{2}\right)$$

◆ **Example 2**
The midpoint of $\overline{AB}$ has coordinates $M(5, 7)$. One endpoint of the segment is $A(3, 4)$. Find the coordinates of point $B(x_2, y_2)$.

◆ **Solution**
Use the midpoint formula to find x_2 and y_2.

$5 = \dfrac{3 + x_2}{2}$ $\qquad\qquad$ $7 = \dfrac{4 + y_2}{2}$

$3 + x_2 = 10$ $\qquad\qquad$ $4 + y_2 = 14$

$x_2 = 7$ $\qquad\qquad\qquad$ $y_2 = 10$

The other endpoint is $B(7, 10)$.

Find the coordinates of the midpoint of each segment with the given endpoints.

5. $A(2, 1)$ and $B(-9, 3)$ $\qquad$ **6.** $C(2, 2)$ and $D(-8, -7)$ $\qquad$ **7.** $E(-1, -3)$ and $F(-3, -5)$

_____   _____   _____

8. The midpoint of a segment has coordinates $M(8, 2)$. One endpoint is $A(6, 0)$.

Find the coordinates of the other endpoint. _____

9. The endpoints of the diameter of a circle have coordinates $R(-8, 2)$ and $S(10, -4)$.

Find the coordinates of the center of the circle. _____

NAME _____ CLASS _____ DATE _____

Reteaching
12.5 Geometric Properties

◆ **Skill A** Finding equations of circles with their centers at the origin

Recall A circle is the set of all points in a plane that are the same distance from a given point called the center. Circle C has its center at the origin, (0, 0). Let r be the length of the radius of the circle and $P(x, y)$ be a point on the circle. Use the distance formula to find the equation of the circle.
$$(x - 0)^2 + (y - 0)^2 = r$$
$$x^2 + y^2 = r^2$$
The equation $x^2 + y^2 = r^2$ represents a circle with its center at the origin and with a radius of r.

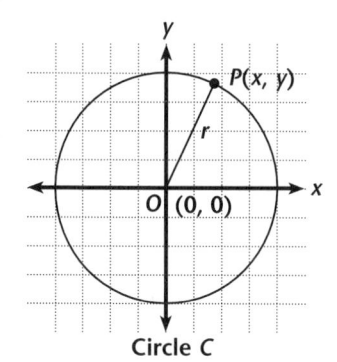

◆ **Example**
Find the equation of the circle with its center at the origin and with a radius of 4.

◆ **Solution**
$x^2 + y^2 = 4^2$
The equation of the circle is $x^2 + y^2 = 16$.

Find the equation of each circle with its center at the origin and with the given radius.

1. radius = 5 feet
2. radius = 2.5 meters
3. radius = $\frac{3}{4}$ inch

_____   _____   _____

◆ **Skill B** Finding the equation of any circle

Recall The equation of circle with center (h, k) and with radius r is $(x - h)^2 + (y - k)^2 = r^2$.

◆ **Example**
Find the center and radius of the circle represented by the equation $(x - 2)^2 + (y + 3)^2 = 25$.

◆ **Solution**
h is 2; k is -3, and $r = \sqrt{25}$, or 5. The circle has its center at $(2, -3)$ and has a radius of 5.

From each equation of a circle, give the center and the radius.

4. $(x + 5)^2 + (y + 1)^2 = 36$
5. $(x - 4)^2 + (y - 1)^2 = 27$
6. $(x - 3)^2 + (y - 3)^2 = 7.84$

_____   _____   _____

Algebra 1   Reteaching Masters   **147**

♦ **Skill C** Using coordinates to check geometric relationships

Recall Geometric figures can be described on a coordinate plane.

♦ **Example**
Check to see if the line segment drawn between the midpoints of the legs of an isosceles triangle is one-half of the length of it base.

♦ **Solution**
The midpoint formula gives (2, 3) for point A and (6, 3) for point B. The distance formula gives the length of $\overline{AB}$ as 4. The length of the base of the triangle is 8. Thus, the segment drawn between the midpoints is one-half of the length of the base.

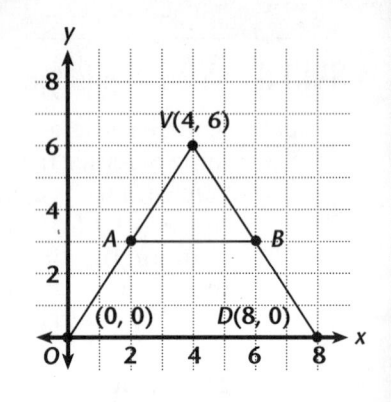

Draw each figure on a coordinate plane. Show that each relationship given is true.

7. A triangle is formed by the points $A(0, 0)$, $B(6, 0)$, and $C(3, 6)$. Show that two sides are equal.

8. A trapezoid is formed by the points $A(0, 0)$, $B(8, 0)$, $C(6, 4)$, and $D(2, 4)$. Show that the lengths of the diagonals are equal.

9. A square is formed by the points $A(0, 0)$, $B(a, 0)$, $C(a, a)$, and $D(0, a)$. Show that the length of a diagonal equals $a\sqrt{2}$.

NAME _____ CLASS _____ DATE _____

Reteaching
12.6 The Tangent Function

◆ **Skill A** Finding tangent ratios

Recall In a right triangle, the tangent of an acute angle is the ratio of the length of the leg opposite the angle to the length of the leg adjacent to the angle.

$\tan \angle A = \dfrac{a}{b}$ $\tan \angle B = \dfrac{b}{a}$

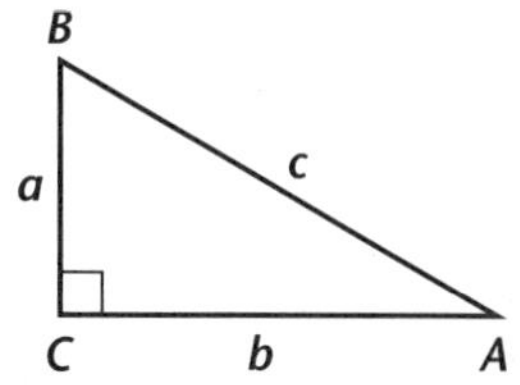

◆ **Example 1**
Find the tangent of $\angle M$ to the nearest thousandth.

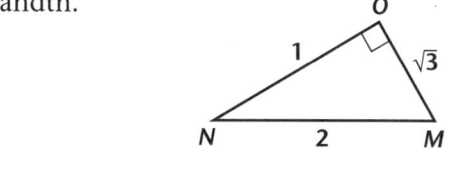

◆ **Solution**

$\tan \angle M = \dfrac{\text{opposite}}{\text{adjacent}} = \dfrac{1}{\sqrt{3}} = \dfrac{\sqrt{3}}{3} \approx 0.577$

The tangent of angle M is about 0.577.

◆ **Example 2**
Use a calculator to find the tangent of 50° to the nearest thousandth.

◆ **Solution**
Use the tangent key on your calculator.
$\tan 50° \approx 1.192$

Find the tangent of each angle X to the nearest thousandth.

1.

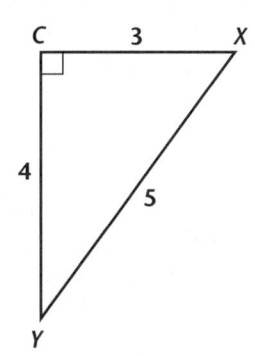

2.

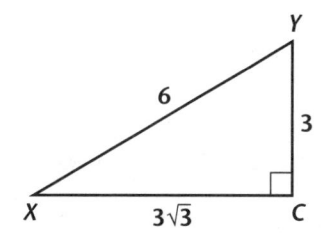

3.

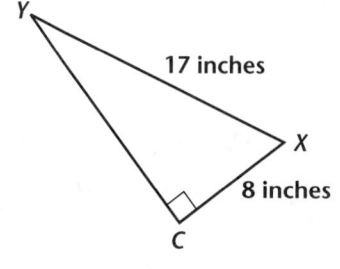

_____   _____   _____

Find the tangent of each angle to the nearest thousandth.

4. 20°   5. 45°   6. 72°

_____   _____   _____

Algebra 1   Reteaching Masters **149**

◆ **Skill B** Using the tangent ratio

Recall The tangent ratio can be used to find a missing dimension in a right triangle.

◆ **Example**
The base of a pole is 20 feet from the stake at point A. The measure of the angle formed by the stake and the top of the pole is 55°. What is the height of the pole?

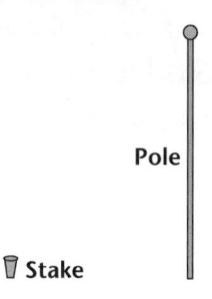

◆ **Solution**
Redraw the diagram and add the given information.
$\tan \angle A = \frac{\text{opp}}{\text{adj}}$
$\tan 55° = \frac{a}{20}$
$a = 20(\tan 55°)$
$a \approx (20)(1.428) \approx 28.56$
The height of the pole is about 28.56 feet.

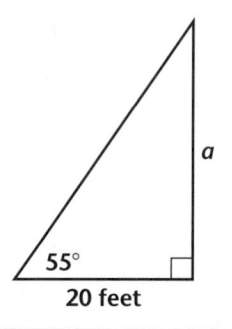

Use the diagrams to find each missing dimension to the nearest tenth.

7. Find the length of $\overline{YZ}$.

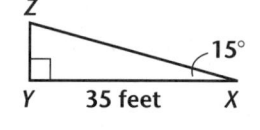

_____

8. The measure of $\angle A$ is 25°. The length of $\overline{AB}$ is 50 meters. Find the length of $\overline{BC}$.

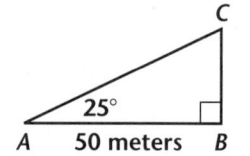

_____

9. The height of a tree is 18 feet. The angle from the marker to the tree is 60°. Find the distance from the tree to the marker.

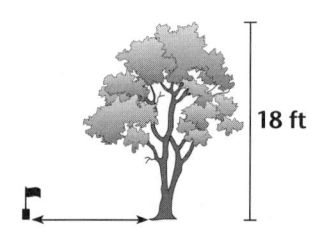

_____

150   **Reteaching Masters**   **Algebra 1**

NAME _____ CLASS _____ DATE _____

Reteaching
12.7 The Sine and Cosine Functions

◆ **Skill A** Finding sines and cosines

Recall The sine and cosine ratios are also based on a right triangle.

Triangle ABC has the following sines and cosines:

$$\sin A = \frac{a}{c} \text{ and } \cos A = \frac{b}{c}$$

$$\sin B = \frac{b}{c} \text{ and } \cos B = \frac{a}{c}$$

◆ **Example**
Find the sine and cosine of $\angle A$.

◆ **Solution**
Because $a = 2$ and $b = 1$, first use the Pythagorean Theorem to find c.
$c = \sqrt{2^2 + 1^2} = \sqrt{5}$

$$\sin A = \frac{2}{\sqrt{5}} = \frac{2\sqrt{5}}{5} \approx 0.894$$

$$\cos A = \frac{1}{\sqrt{5}} = \frac{\sqrt{5}}{5} \approx 0.447$$

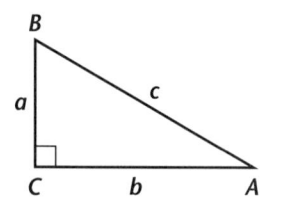

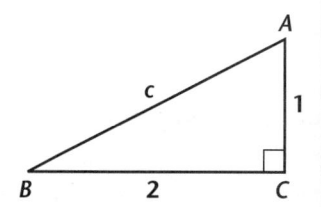

Find the sine and cosine of each $\angle A$. Round to the nearest thousandth, if necessary.

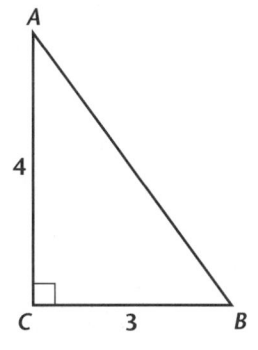

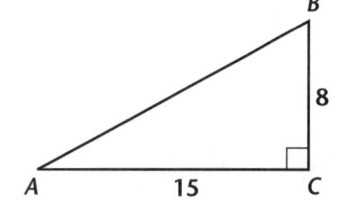

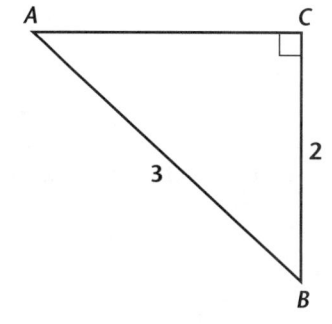

1. (triangle with legs 4 and 3, right angle at C)

2. (triangle with legs 15 and 8, right angle at C)

3. (triangle with legs 3 and 2, right angle at C)

_____   _____   _____

Find the sine and cosine of each angle to the nearest thousandth.

4. 25°   5. 45°   6. 90°

_____   _____   _____

Algebra 1   Reteaching Masters   **151**

NAME _____ CLASS _____ DATE _____

◆ **Skill B** Using sine and cosine

Recall The sine and cosine ratios can be used to find missing dimensions of a right triangle.

◆ **Example**
A 6-foot ramp is placed on the front end of a step. The ramp makes a 40° angle with the step. Find the height of the step.

◆ **Solution**
Draw a diagram to show the information.

$\sin A = \dfrac{a}{c}$

$\sin 40° = \dfrac{a}{6}$

$a = 6(\sin 40°)$
$a \approx (6)(0.643) \approx 3.86$

The height of the step is about 3.86 feet.

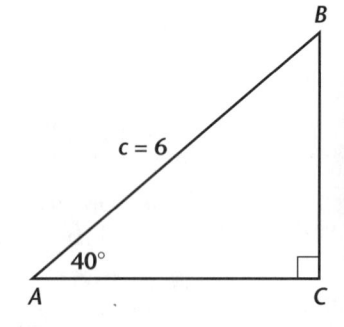

Use a diagram to help find each missing dimension to the nearest tenth.

7. A kite is on a string that is 50 feet long. The string forms an angle of 35° with the ground. How high above the ground is the kite? _____

8. Two miles of a railroad track is on a grade that forms a 15° angle with the horizontal. A 1500-meter train is on the track. How much higher is the front of the train than the rear of the train? _____

9. A marker is placed 300 feet from the base of a hill. The marker makes a 20° angle with the top of the hill. What is the distance from the marker to the top of the hill? _____

152   **Reteaching Masters**   Algebra 1

NAME _____ CLASS _____ DATE _____

Reteaching
12.8 Introduction to Matrices

◆ **Skill A** Adding and subtracting matrices

Recall Matrices can be added only if they have the same dimensions. If they do, then add the corresponding entries.

◆ **Example 1**
Let $A = \begin{bmatrix} -5 & 3 \\ 2.5 & -1 \end{bmatrix}$ and let $B = \begin{bmatrix} 2.6 & -4.5 \\ -2.5 & 6 \end{bmatrix}$. Find $A + B$.

◆ **Solution**
$A + B = \begin{bmatrix} -5 + 2.6 & 3 + (-4.5) \\ 2.5 + (-2.5) & -1 + 6 \end{bmatrix} = \begin{bmatrix} -2.4 & -1.5 \\ 0 & 5 \end{bmatrix}$

Recall Matrices can be subtracted only if they have the same dimensions. If matrices A and B have the same dimensions, you can calculate $A - B$ by finding the opposite of B and then adding that matrix to A.

◆ **Example 2**
Let $A = \begin{bmatrix} 3 & -1 & -2.7 \\ 5 & 1.4 & -5 \end{bmatrix}$ and let $B = \begin{bmatrix} 0.6 & -1 & 0 \\ 2.6 & 4 & -5 \end{bmatrix}$. Find $A - B$.

◆ **Solution**
The opposite of B is $\begin{bmatrix} -0.6 & 1 & 0 \\ -2.6 & -4 & 5 \end{bmatrix}$.

Add the opposite of B to A.
$ A + (-B)$
$\begin{bmatrix} 3 & -1 & -2.7 \\ 5 & 1.4 & -5 \end{bmatrix} + \begin{bmatrix} -0.6 & 1 & 0 \\ -2.6 & -4 & 5 \end{bmatrix} = \begin{bmatrix} 2.4 & 0 & -2.7 \\ 2.4 & -2.6 & 0 \end{bmatrix}$

Perform the indicated matrix operations. If a solution is not possible, explain why.

Let $A = \begin{bmatrix} 3.2 & -5.4 & 8 \\ 2.7 & 1.5 & -3 \end{bmatrix}$, $B = \begin{bmatrix} 1.5 & -2.5 & 7 \\ -3.5 & -1.5 & 4 \end{bmatrix}$, $C = \begin{bmatrix} 4 & 0 \\ -2 & 6 \\ 0 & -1 \end{bmatrix}$, and $D = \begin{bmatrix} -1.5 & 3.5 \\ 2.5 & 1.5 \\ 7 & -4 \end{bmatrix}$.

1. $A + B$ _____

2. $B + C$ _____

3. $C + D$ _____

4. $B - A$ _____

5. $A - D$ _____

6. $D - C$ _____

Algebra 1   Reteaching Masters   **153**

◆ **Skill B** Multiplying matrices

Recall When you multiply a matrix by a number, you form the scalar product of the matrix and number. To perform this multiplication, multiply each entry in the matrix by the scalar.

◆ **Example 1**

Find $-3\begin{bmatrix} 0 & -2 & 3 \\ 1 & 4 & 5 \\ -3 & -5 & 2 \end{bmatrix}$.

◆ **Solution**

$$-3\begin{bmatrix} 0 & -2 & 3 \\ 1 & 4 & 5 \\ -3 & -5 & 2 \end{bmatrix} = \begin{bmatrix} (-3)(0) & (-3)(-2) & (-3)(3) \\ (-3)(1) & (-3)(4) & (-3)(5) \\ (-3)(-3) & (-3)(-5) & (-3)(2) \end{bmatrix} = \begin{bmatrix} 0 & 6 & -9 \\ -3 & -12 & -15 \\ 9 & 15 & -6 \end{bmatrix}$$

Thus, $-3\begin{bmatrix} 0 & -2 & 3 \\ 1 & 4 & 5 \\ -3 & -5 & 2 \end{bmatrix} = \begin{bmatrix} 0 & 6 & -9 \\ -3 & -12 & -15 \\ 9 & 15 & -6 \end{bmatrix}$.

Recall If A is an $m \times n$ matrix and B is an $n \times p$ matrix, then you can find the matrix product, AB. To find the product, use the rows of matrix A and the columns of matrix B.

◆ **Example 2**

Let $A = \begin{bmatrix} 2.5 & 1 \\ 0 & -3 \end{bmatrix}$ and $B = \begin{bmatrix} 3 & 1.5 \\ 0 & -4 \end{bmatrix}$. Does AB exist? If so, find AB.

◆ **Solution**

Since the number of columns in A equals the number of rows in B, AB exists. To find the entry in row 1 and column 1 of the product, use row 1 of A and column 1 of B.

$$[2.5 \quad 1] \text{ and } \begin{bmatrix} 3 \\ 0 \end{bmatrix} \longrightarrow 2.5 \times 3 + 1 \times 0 = 7.5$$

To find all the entries in AB, use all the rows of A and all the columns of B.

$$AB = \begin{bmatrix} (2.5)(3) + (2)(0) & (2.5)(1.5) + (1)(-4) \\ (0)(3)(-3)(0) & (0)(1.5) + (-3)(-4) \end{bmatrix} = \begin{bmatrix} 7.5 & 3.75 \\ 0 & 12 \end{bmatrix}$$

Let $A = \begin{bmatrix} -3 & 0 \\ -7 & 5 \end{bmatrix}$, $B = \begin{bmatrix} 0 & -1 & 1 \\ 2 & -2 & 0 \\ 5 & -5 & 0 \end{bmatrix}$, $C = \begin{bmatrix} 3 & 3 & -1 \\ 0 & 5 & -7 \end{bmatrix}$, and $D = \begin{bmatrix} 12 & -12 \\ -12 & 12 \end{bmatrix}$.

Find each product if it exists. If it does not, explain why not.

7. AB 8. CB 9. AD 10. AC

Reteaching
13.1 Theoretical Probability

◆ **Skill A** Listing favorable outcomes

Recall The set of all possible outcomes of a probability experiment is called the sample space of the experiment.

◆ **Example 1**
What is the sample space when a coin is tossed three times?

◆ **Solution**
There are eight different possible outcomes.

First Toss	Second Toss	Third Toss
H	H	H
H	H	T
H	T	H
H	T	T
T	H	H
T	H	T
T	T	H
T	T	T

The sample space is HHH, HHT, HTH, HTT, THH, THT, TTH, TTT.

Recall The favorable outcomes are the outcomes that you are looking for in the probability experiment.

◆ **Example 2**
What are the favorable outcomes if you want exactly two heads to appear?

◆ **Solution**
The favorable outcomes are HHT and THH.

A coin is tossed, and a number cube is rolled.

1. List the sample space. _____

2. What are the favorable outcomes if you want a head OR a 2 to appear? _____

3. What are the favorable outcomes if you want a head AND a 4 to appear? _____

4. What are the favorable outcomes if you want a tail OR an even number to appear? _____

5. What are the favorable outcomes if you want a tail AND an even number to appear? _____

Algebra 1   Reteaching Masters **155**

NAME _____ CLASS _____ DATE _____

◆ **Skill B** Finding the probability that an event will occur

Recall The theoretical probability, P, that an event will occur is defined as $P = \frac{f}{n}$, where f is the number of favorable outcomes and n is the total number of equally likely outcomes.

◆ **Example**
Find the probability that in one roll of a 12-sided number cube (dodecahedron) a prime number will result.

◆ **Solution**
Each of the 12 faces is equally likely to land face up on any one roll. Because 2, 3, 5, 7, and 11 are the prime numbers between 1 and 12, there are 5 ways for a successful outcome to occur.

$$P = \frac{\text{number of favorable outcomes}}{\text{total number of outcomes}} = \frac{f}{n} = \frac{5}{12}$$

The probability of rolling a prime number is $\frac{5}{12}$.

Find each probability.

6. From a bag containing 12 marbles, one is drawn at random. If the bag contains 2 yellow, 3 green, 4 blue, and 3 red marbles, find the probability of drawing a find the probability of drawing a green marble. _____

7. If a letter is selected at random from the word *probability,* find the probability that the letter is a consonant. _____

8. If a letter is selected at random from all the letters in this sentence, find the probability that the letter selected is a *t*. _____

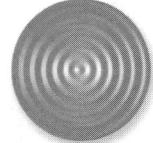

Reteaching
13.2 Counting the Elements of Sets

◆**Skill A** Drawing and using Venn diagrams

Recall In a Venn diagram, overlapping regions represent the intersection of two sets, or the word AND. Combined regions represent the union of two sets, or the word OR. The area outside a region represents the word NOT.

◆ **Example**
In a small Ghanaian village, there are 38 children with ages ranging from 6 to 12 years old. Eighteen of the children eat kenkey and 27 eat fufu. This includes 7 children who eat both. Use a Venn diagram to determine how many children eat only fufu.

◆ **Solution**
Because some children eat both foods, draw two circles that overlap. In the overlap, place the number that represents the number of children who eat both foods.

The total number of children who eat kenkey must be 18, and 7 of the children that eat kenkey are already represented, so place an additional 11 in the other part of the kenkey circle. The total number of children who eat fufu is 27. Because 7 that eat fufu are already represented, an additional 20 must be placed within the other part of the fufu circle. Thus, there are 20 children who eat only fufu.

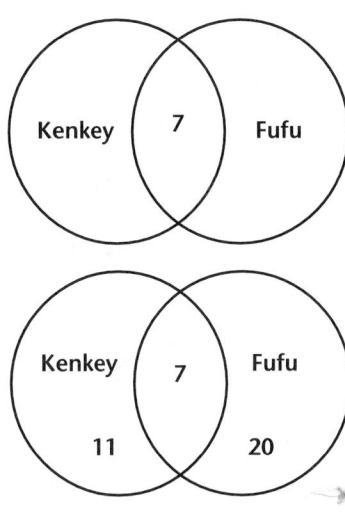

Represent each situation with a Venn diagram.

1. Fifteen children at a birthday party are taken to an ice-cream parlor. Seven order chocolate, and 12 order vanilla. This includes 4 who order a combination of chocolate and vanilla.

2. At a summer camp, 36 campers were asked whether they play golf or tennis. Fifteen said that they play golf, while 17 reported that they play tennis. This included 2 who play both and 6 who play neither.

Algebra 1   Reteaching Masters **157**

◆ **Skill B** Counting elements of sets

Recall The total number of ways to choose A OR B = (number of ways to chose A + number of ways to choose B) − (number of ways to choose both A AND B).

◆ **Example**

A survey of students' preferences for spring sports produced the following results:

	Baseball	Lacrosse	Total
Blue	14	16	30
Orange	12	19	31
Total	26	35	61

a. How many students were in the orange group?
b. How many students preferred baseball?
c. How many students were in the orange group AND preferred baseball?
d. How many students were in the orange group OR preferred baseball?

◆ **Solution**

a. The total for the second row, orange, is 31.
b. The total for the first column, baseball, is 26.
c. The intersection of the first column, baseball, and the second row, orange, is 12.
d. Number of students who were in the orange group OR preferred baseball =

     Number of students who were in orange group   31
  + Number of students who preferred baseball   + 26
                                                          57
  − Number of students who were in orange group AND   − 12
     preferred baseball   45 students

The Parents' Sports Club conducted a survey in order to determine whether to hold an auction or to run a raffle to raise funds. The table shows the results of the survey.

	Auction	raffle	Total
Men	44	91	135
Women	116	35	151
Total	160	126	286

3. How many parents surveyed were men? _____

4. How many parents preferred a raffle? _____

5. How many parents surveyed were men AND preferred a raffle? _____

6. How many parents surveyed were men OR preferred a raffle? _____

Reteaching
13.3 The Fundamental Counting Principle

◆ **Skill A** Drawing and using tree diagrams

Recall Each branch of a tree diagram represents a possible outcome. The number of branches at each endpoint depends on the number of ways in which an event can happen. Together, all of the branches represent all of the possible outcomes of an event.

◆ **Example**
Three flavors of ice cream, four sauces, and two toppings, either cherries or chocolate sprinkles, are available for making sundaes. Use a tree diagram to determine how many different sundaes can be made with one flavor of ice cream, one sauce, and one topping.

◆ **Solution**
Start with one point and make three branches to represent the three flavors of ice cream.

From each point at the end of these three branches, make four branches to represent the four sauces.

From each point at the end of the 12 branches, make two branches to represent the two toppings.

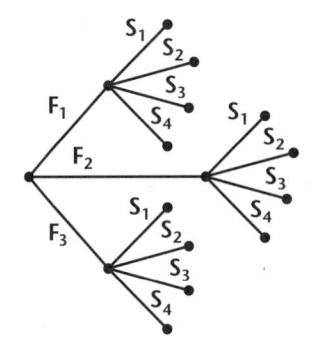

Count the number of paths in the tree. There are 24 different sundaes that can be made with the given ingredients.

An alternative is to count the points at the ends of the paths.

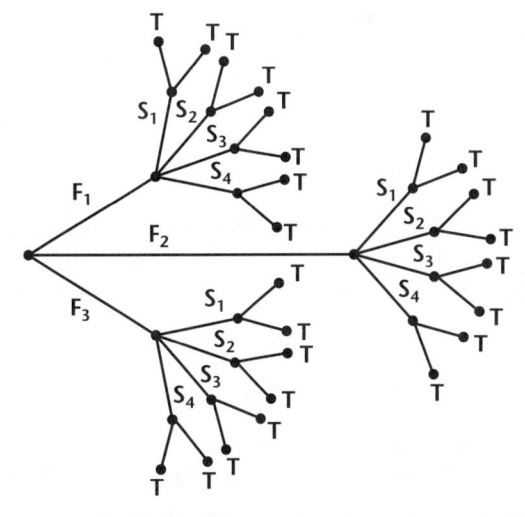

Use a tree diagram to answer each question.

1. When packing for a business trip, Mr. Simmons selects 3 suits, 4 shirts, and 5 ties. How many different outfits (1 suit, 1 shirt, 1 tie) can he put together? _____

Algebra 1   Reteaching Masters

2. The science class decided to present a pet display for the kindergarteners. Two students volunteered to bring fish; 3 could bring cats; 3 wanted to bring dogs, and 2 said that they could bring rabbits. If the display could include one pet from each category, how many displays were possible? _____

◆ **Skill B**   Using the Fundamental Counting Principle

Recall   The total number of ways to choose A AND B AND C = number of ways to choose A · number of ways to choose B · number of ways to choose C.

◆ **Example**
When Marianne got a new job, she bought 5 blouses, 4 skirts, and 3 pairs of shoes. How many days could Marianne go to work without wearing the same outfit (1 blouse, 1 skirt, 1 pair of shoes)?

◆ **Solution**
To put an outfit together, there are 5 tops to choose from. Once a top is chosen, there are 4 skirts to choose from. Once a top and skirt are chosen, there are 3 pairs of shoes to choose from.

Number of outfits =
number of ways to choose a top · number of ways to choose a skirt · number of ways to choose a pair of shoes 5 · 4 · 3 = 60 outfits

---

Use the Fundamental Counting Principle to solve each of the following:

3. If the new license plate pattern in the 51st state is any letter, followed by any digit, followed by any letter, how many distinct license plates can be made? _____

4. The house numbers in a new development are to be 3-digit numbers ranging from 700 to 899 and the last digit must be even. How many possible house numbers will there be? _____

5. Two cards are drawn from a regular deck of 52 playing cards one at a time without replacement. How many ways are there to draw 2 kings? _____

NAME _____ CLASS _____ DATE _____

Reteaching
13.4 Independent Events

◆ **Skill A** Defining independent events

Recall In independent events, the occurrence of the first event does not affect the probability that the second event will occur.

◆ **Example**
Determine whether the following are independent:

a. A number cube is rolled, and a coin is tossed. What is the probability that the number cube displays an odd number AND the coin displays tails?

b. Two cards are drawn from a standard deck without replacement (that is, the first card is not put back before the second is drawn). What is the probability that both cards are kings?

◆ **Solution**
a. Tossing a coin has no relationship to rolling a number cube. Thus, the probability of the coin displaying tails is not affected by the outcome of the roll of the number cube. The events are independent.

b. Of the 52 cards in a standard deck, 4 are kings, so the probability of drawing a king is $\frac{4}{52}$, or $\frac{1}{13}$. If a king is drawn first and not replaced, there are 3 kings remaining out of 51 cards, so on the second draw, the probability of drawing a king is $\frac{3}{51}$, or $\frac{1}{17}$. Thus, the probability of the second event (drawing a king) depends on the occurrence of the first event (drawing a king). The events are not independent.

Determine whether each of the following are independent events.

1. Two marbles are drawn from a bag without replacement. If the bag contains 4 green, 5 red, and 3 blue marbles, what is the probability of drawing 2 red marbles? _____

2. A card is drawn from a standard 52-card deck, and a number cube is rolled. What is the probability of drawing a queen AND rolling a 5? _____

3. Two number cubes are rolled. What is the probability that the sum of the numbers rolled is 9 AND the second cube shows a 3? _____

4. A card is drawn from a standard 52-card deck and replaced. A second card is then drawn. Find the probability that 2 aces are drawn. _____

Algebra 1   Reteaching Masters   **161**

◆ **Skill B** Finding the probability of independent events

Recall If events A and B are independent, then the probability of A AND B = probability of A · probability of B.

◆ **Example**
Andre and Angela play darts as a team against other two-person teams. If Andre makes a bull's-eye 70% of the time and Angela makes a bull's-eye 60% of the time, find the probability that both will hit a bull's-eye when making one throw apiece.

◆ **Solution**
The probability that Andre makes a bull's-eye on any shot is 70% = $\frac{70}{100} = \frac{7}{10}$.

The probability that Angela makes a bull's-eye on any shot is 60% = $\frac{60}{100} = \frac{6}{10}$.

Since the events are independent, the probability that both Andre and Angela make a bull's-eye is $\frac{7}{10} \cdot \frac{6}{10} = \frac{42}{100} = 42\%$.

This can be pictured on a 10 × 10 square grid, where 70 of the squares represent Andre's probability of making a bull's-eye and 60 of the squares represent Angela's probability. These regions overlap on 42 squares, which represents the probability that both make a bull's-eye.

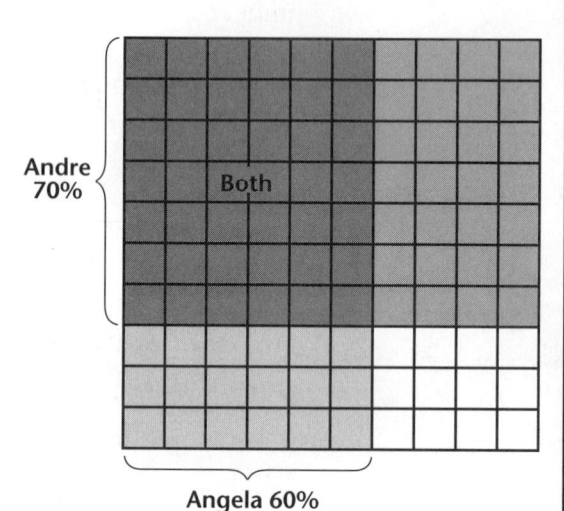

Find the probability of each situation described.

5. Two number cubes are rolled. Find the probability that both cubes show a 3 or less.

6. Two cards are drawn from a standard 52-card deck. The first is replaced before the second is drawn. Find the probability that both cards drawn are spades.

7. A coin is flipped, and a number cube is rolled. Find the probability that the coin shows heads AND the cube shows a 6.

8. The digits from 0 to 9 are printed on 10 Ping-Pong balls and are placed in a bag. Two balls are drawn, with the first ball being replaced before the second is drawn. What is the probability of drawing the digits 7 AND 5, in that order, on the two draws?

162   Reteaching Masters   Algebra 1

NAME _____ CLASS _____ DATE _____

Reteaching
13.5 Simulations

◆ **Skill A** Using coins as simulation devices

Recall Because a coin toss has only two possible outcomes—heads or tails—it is the random generator to use when the situation being simulated has exactly two outcomes.

◆ **Example**
Chris and Paul play checkers every day. Before starting, Paul draws without looking from a box containing one red piece and one black piece in order to decide which color he will play. In the month of September, how many days is Paul likely to choose red? What is the experimental probability?

◆ **Solution**
Since there are exactly two outcomes, red and black, a coin can be used to simulate the outcomes.

Let heads represent red and let tails represent black. Let each toss represent one daily draw. Because the problem specifies the month of September, which has 30 days, 30 trials should be generated, and the results should be recorded.

Sample set of outcomes:

THTHTTTHTH
HHHTHHTTTH
THHTTTHHHT

In this simulation, there are 15 heads. This represents 15 days in September that Paul is likely to choose red. Thus, the experimental probability of Paul choosing red is $\frac{15}{30} = \frac{1}{2}$, or 50%.

Using a coin as a random generator, design and perform a simulation for each situation described.

1. A teacher asked 24 students to guess the answer to this question: "I have a quarter in my pocket—true or false?" If the teacher did not have a quarter, how many students guessed correctly?

_____

_____

2. As part of her weekly workout, Arlene does sit-ups and push-ups. To decide which she should do first, she looks at her digital watch. If the seconds indicator is an even number, she does sit-ups first. What is the probability that she starts her workout with sit-ups?

_____

_____

Algebra 1   Reteaching Masters **163**

NAME _____ CLASS _____ DATE _____

◆ **Skill B** Designing simulations

Recall Random values can be generated by rolling a number cube, tossing a coin, spinning a spinner, or using a calculator.

◆ **Example**
Design a simulation to find the experimental probability for the following situation:

The weather forecaster predicts a 50% chance of rain on Saturday and an 80% chance of rain on Sunday. What is the experimental probability that it will rain on both days?

◆ **Solution**
Decide what type of random number generator to use for the simulation. Number cubes of different colors could be used for both predictions. The numbers 1, 2, and 3 on one cube could represent rain on Saturday while 4, 5, and 6 could represent no rain. On the second cube, the numbers 1, 2, 3, and 4 could represent rain on Sunday, while 5 could represent no rain. A roll of 6 would mean that you roll again. Or a coin could be flipped for Saturday's results, using heads for rain and tails for no rain.

Each pair of results would represent one trial, or outcome. Generate 20 trials and record the results.

Design a simulation to find the experimental probability for each situation described.

3. If a playing card is drawn at random from a standard 52-card deck, what is the probability that it will be a red card? _____

   _____

4. A candy dish contains 60 candies, 10 of each of six colors: red, green, yellow, blue, tan, and dark brown. If Paula takes one candy from the dish without looking, what is the probability that she will pick a tan candy? _____

   _____

5. A 40% chance of rain on Saturday and a 50% chance of rain on Sunday are predicted. What is the probability that it will not rain on either day?

   _____

   _____

Reteaching
14.1 Graphing Functions and Relations

◆ **Skill A**   Understanding differences between relations and functions

Recall   A relation is a set of ordered pairs.
A function is a relation in which each x-coordinate is paired with exactly one y-coordinate.

◆ **Example**
State whether each set of ordered pairs is a function.
  a. $\{(3, 1), (4, 6), (0, 1), (3, 2), (-4, 6), (5, 2)\}$
  b. $\{(1, 0), (-5, 9), (4, 6), (2, 3), (5, 8), (-2, 4), (0, 4)\}$

◆ **Solution**
  a. If all the x-coordinates are different, the set is a function. If any x-coordinates repeat, check their y-coordinates. Because $(3, 1)$ and $(3, 2)$ have the same x-coordinate but different y-coordinates, the relation is not a function.
  b. Because all the x-coordinates are different, the set is a function.

Decide whether each set is a function.

1. $\{(3, 5), (2, 6), (4, 6), (-1, 6), (5, 8)\}$ _____
2. $\{(1, 0), (1, 4), (1, -3)\}$ _____
3. $\{(-6, 8), (6, 8), (-5, 7), (5, 7)\}$ _____
4. $\{(5, 1), (4, 2), (3, 3), (2, 4)\}$ _____

◆ **Skill B**   Using the $f(x)$ function notation to represent and evaluate functions

Recall
$$f(x) = ax^2 + bx + c$$
↑
replacement variable   function rule

Substitute the value of the replacement variable into the function rule and simplify.

◆ **Example**
Let $f(x) = 5x^2 - 3x + 2$. Find $f(3)$.

◆ **Solution**
Replace each x in the function rule with a 3.
$f(3) = 5(3)^2 - 3(3) + 2$
$= 5(9) - 3(3) + 2$
$= 45 - 9 + 2$
$= 38$
Thus, $f(3) = 38$.

Evaluate each function.

5. Let $f(x) = 7x^2 - 5x$. Find $f(-2)$. _____
6. Let $g(x) = -x^2 + 2x$. Find $g(4)$. _____

Algebra 1   Reteaching Masters   165

NAME _____ CLASS _____ DATE _____

Evaluate each function when x is −3.

7. $f(x) = -2x^2 - x$ _____ **8.** $g(x) = \dfrac{1}{x^2}$ _____

◆ **Skill C** Using the vertical-line test to identify functions

Recall Two points whose x-coordinates are the same but whose y-coordinates are different lie on the same vertical line.

◆ **Example**
Use the vertical-line test to decide whether each graph represents a function.

a.

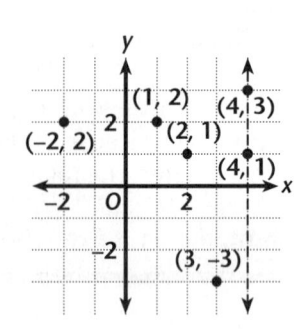

b.

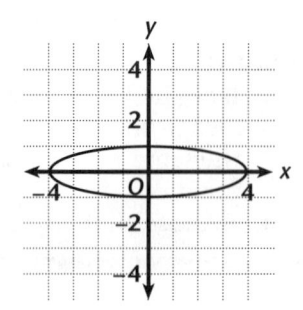

◆ **Solution**
  a. The relation shown has six points. Two of these points lie on the vertical line $x = 4$. Therefore, the graph does not represent a function.
  b. Because any vertical line drawn between $x = -4$ and $x = 4$ will cross the graph at more than one point, the graph does not represent a function.

Use the vertical-line test to decide whether each graph represents a function.

9.

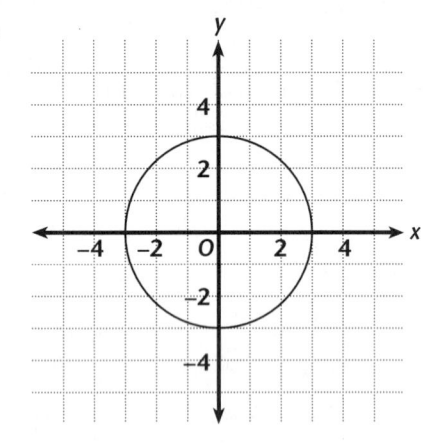

10.

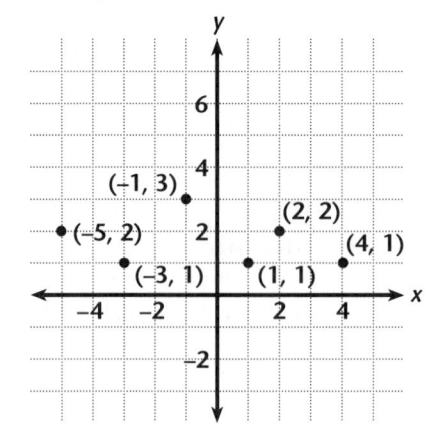

---

166   **Reteaching Masters**   **Algebra 1**

NAME _____ CLASS _____ DATE _____

Reteaching
14.2 Translations

◆ **Skill A** Describing the effects of translations on the graphs of functions

Recall When a function is translated vertically or horizontally, every point is moved the same distance in the same direction.

◆ **Example 1**
One point on the graph of $y = |x| + 3$ is $(-4, 7)$. What is the resulting point if the function is translated vertically by -4?

◆ **Solution**
Since every point is moved the same distance in the same direction, the point is moved down 4 units. A vertical change affects only the y-value of a point, so the new y-value is $7 - 4$, or 3. Therefore, the resulting point is $(-4, 3)$.

◆ **Example 2**
Choose a point on the graph of $y = x^2 - 2$. Then find the corresponding point after the function is translated horizontally by 3.

◆ **Solution**
Choose any value to substitute for x in the equation. If $x = 4$, for example, $y = 4^2 - 2$, or 14. Therefore, the coordinates of the point are $(4, 14)$. Because a horizontal translation affects only the x-coordinate of a point, 3 must be added to the first coordinate of the point $(4, 14)$. The new value of x is $4 + 3$, or 7. The coordinates of the point corresponding to $(4, 14)$ after the horizontal translation are $(7, 14)$.

Using the function $y = |x| + 4$, find each point after the indicated translation.

1. $(-2, 6)$, vertical by -2 _____

2. $(5, 9)$, horizontal by 4 _____

3. $(-7, 11)$, horizontal by -3 _____

4. $(3, 7)$, vertical by -5 _____

5. $(13, 17)$, vertical by 6 _____

6. $(-5, 9)$, horizontal by -12 _____

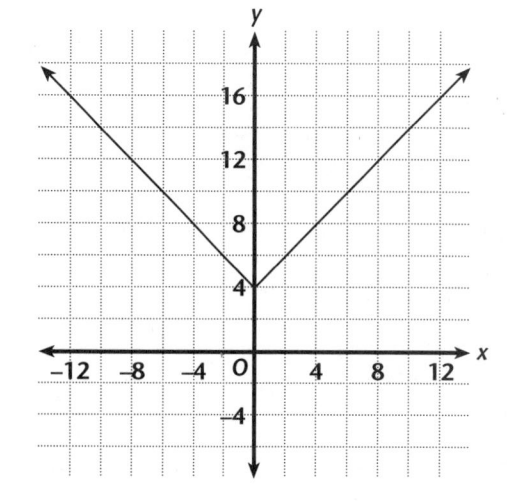

Algebra 1   **Reteaching Masters**   **167**

NAME _____ CLASS _____ DATE _____

◆ Skill B Identifying relationships between the translation of a graph and the addition or subtraction of a constant

Recall A horizontal translation of the parent function $y = f(x)$ occurs when a constant is added to or subtracted from the variable x. Adding 2 to the variable in the function x^2 results in $(x + 2)^2$, which shifts the parent function 2 units to the left.

A vertical translation of the parent function $y = f(x)$ occurs when a constant is added to or subtracted from the function value $f(x)$. Adding 2 to the function x^2 results in $x^2 + 2$, which shifts the parent function 2 units up.

◆ Example
Identify the parent function. Describe the effect of the addition or subtraction on the parent function.

a. $y = x^2 + 4$ **b.** $y = |x + 5|$

◆ Solution
a. The parent function is $y = x^2$. Once each x-value is squared, the result is then increased by 4. The effect of this addition on the parent function is a shift 4 units up.
b. The parent function is the absolute-value function, $y = |x|$. Before taking the absolute value of each x, the x-value is increased by 5 or decreased by -5. The effect of this subtraction on the parent function is a shift 5 units to the left.

Identify each parent function. Describe the effect of the addition or subtraction on the parent function.

7. $y = 2^x + 3$ _____

8. $y = \frac{1}{x} - 2$ _____

9. $y = x^2 + 2$ _____

10. $y = |x| - 4$ _____

11. $y = 10^{(x-2)}$ _____

12. $y = |x + 3|$ _____

13. $y = (x - 4)^2$ _____

14. $y = x^2 - 3$ _____

NAME _____ CLASS _____ DATE _____

teaching

.3 Stretches

cribing the effects of stretches on the graphs of functions

ltiplying a function by a scale factor greater than 1 produces function values t are larger than those of the parent function; multiplying by a scale factor ween 0 and 1 produces function values that are smaller than those of the parent ction.

Example 1
dentify the parent function and the scale factor for $y = 6|x|$. Describe the effect of the stretch and sketch the graph of the transformed function.

Solution
The parent function is the absolute-value function, $y = |x|$. The scale factor is 6, the coefficient of $|x|$. Because the scale factor is greater than 1, the y-values of the stretched function will be 6 times larger than the y-values of the parent function. The graph of the function will rise or fall more steeply as it is stretched vertically. First sketch the graph of the parent function. Recall that it has a V-shape and that its two branches extend through the first and second quadrants.

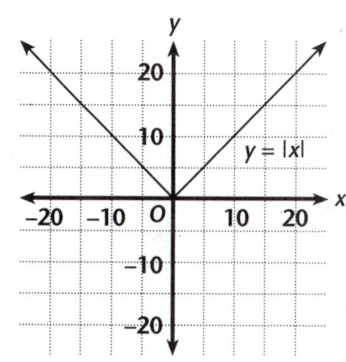

Next choose a point on the parent function. Then apply the scale factor to determine a point on the stretched graph. For example, (2, 2) becomes (2, 12). Sketch the resulting graph.

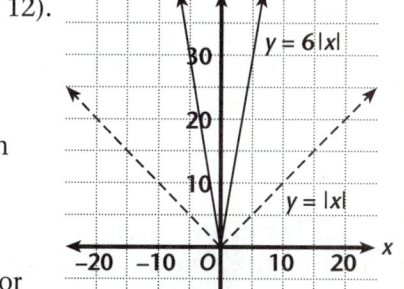

◆ **Example 2**
Describe the effect of the stretch on the function $y = \frac{|x|}{5}$.

◆ **Solution**
The parent function is $y = |x|$, and the scale factor is $\frac{1}{5}$. The y-values of the stretched function will be $\frac{1}{5}$ of the y-values of the parent function. The V-shape of the graph will rise or fall less steeply than that of the parent function.

Identify the parent function and scale factor of each function.

1. $y = \frac{3}{x}$ _____
2. $y = \frac{x^2}{5}$ _____
3. $y = 3|x|$ _____
4. $y = 2(10^x)$ _____

Describe the effect of the stretch on each function and sketch the resulting graph on graph paper.

5. $y = 3 \cdot 2^x$ _____
6. $y = \frac{x^2}{4}$ _____

Algebra 1   Reteaching Masters

NAME _____ CLASS _____ DATE _____

◆ **Skill B** Identifying coefficients in order to determine the amount of stretch on the graph of a function

Recall Division can be expressed as multiplication by the reciprocal: $\frac{x}{a} = \frac{1}{a} \cdot x$.

◆ **Example**
For each function, identify the coefficient. Then determine the amount and direction of stretch produced by the scale factor.

a. $y = \frac{1}{4x}$ b. $y = \frac{4}{x}$

◆ **Solution**
a. The function can be rewritten as $y = \frac{1}{4} \cdot \frac{1}{x}$. The coefficient is $\frac{1}{4}$. Every y-value of the parent function is multiplied by $\frac{1}{4}$, so each point of the parent function, $y = \frac{1}{x}$, is moved to $\frac{1}{4}$ of the distance from the x-axis.

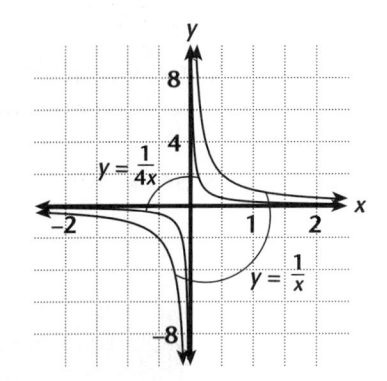

b. The function can be rewritten as $y = 4 \cdot \frac{1}{x}$, so the coefficient is 4. Each y-value of the parent function is multiplied by 4. The positive y-values, which lie in the first quadrant, are 4 times their original values. The negative y-values, which are in the third quadrant, are also 4 times their original values. In the resulting graph, each point is 4 times the distance from the x-axis.

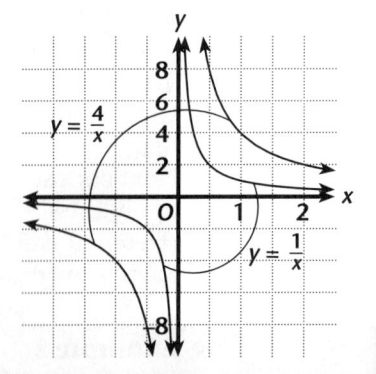

For each function, identify the coefficient. Then determine the amount of the stretch produced by the scale factor.

7. $y = 2x$ _____

8. $y = 3|x|$ _____

9. $y = \frac{5}{x}$ _____

10. $y = \frac{|x|}{3}$ _____

11. $y = \frac{x}{4}$ _____

12. $y = \frac{x^2}{5}$ _____

13. $y = \frac{2|x|}{5}$ _____

14. $y = 0.65|x|$ _____

170   Reteaching Masters   Algebra 1

# Reteaching
14.4 Reflections

◆ **Skill A** Determining if the image of a function is a vertical reflection of its parent function through the x-axis

Recall Points (a, b) and $(a, -b)$ are the same distance from the x-axis along a line perpendicular to the x-axis. These points are vertical reflections of each other through the x-axis.

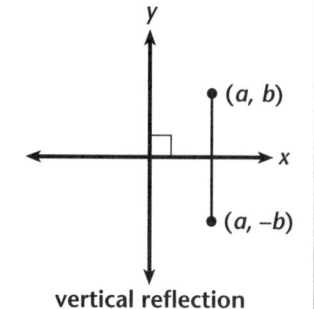

◆ **Example 1**
Determine whether $y = -x^2$ is a vertical reflection of the parent function $y = x^2$.

◆ **Solution**
In a vertical reflection, every point (a, b) of the function is matched with its reflection point $(a, -b)$. Compare the table of values for the two functions.

x	-2	-1	0	1	2
x^2	4	1	0	1	4

x	-2	-1	0	1	2
$-x^2$	-4	-1	0	-1	-4

Because $-x^2 = (-1)x^2$, each y-value of the second function is the opposite of each y-value of the first function. Every point (a, b) on $y = x^2$ is matched by its reflection point $(a, -b)$ on $y = -x^2$. Thus, $y = -x^2$ is a vertical reflection of the parent function.

Recall Two points are vertical reflections if the y-coordinates of the same x-coordinates are opposites.

◆ **Example 2**
Determine whether $y = |x| - 1$ is a vertical reflection of the parent function $y = |x|$.

◆ **Solution**
If the function $y = |x| - 1$ is the vertical reflection of the parent function $y = |x|$, each point (a, b) of the first function will be matched with its reflection point $(a, -b)$. Choose an x-value other than 0 or 1 as a test value.
For $x = 3$,
$y = |x| - 1$ $y = |x|$
$y = 2$ $y = 3$
Since (3, 2) and (3, 3) do not have the same y-coordinate, the function $y = |x|$ is not a vertical reflection of $y = |x|$.

Algebra 1   Reteaching Masters

NAME _____ CLASS _____ DATE _____

Determine whether the given function is a vertical reflection of its parent function.

1. $y = -|x|$

2. $y = 3x - 2$

3. $y = 2^{-x}$

4. $y = x^2 - 2$

5. $y = -10^x$

6. $y = \dfrac{-1}{x}$

◆ **Skill B** Graphing a vertical reflection of a function

Recall To graph a vertical reflection, draw the vertical reflection of each point. Each point on the reflection will be the same distance from the x-axis but located on the other side of the x-axis.

◆ **Example**
Explain how to draw the vertical reflection of the graph of at right.

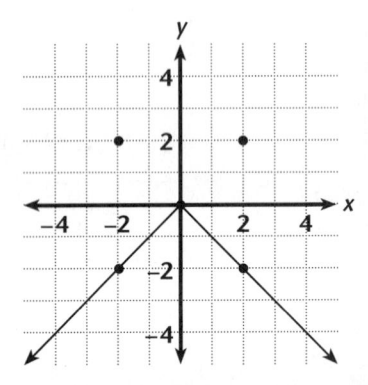

◆ **Solution**
Choose three points, such as $(-2, 2)$, $(0, 0)$, and $(2, -2)$, on the graph. Use the same x-coordinates and the opposite of each y-coordinate to plot three points of the new image. Thus, the vertical reflection will pass through the points $(-2, 2)$, $(0, 0)$, and $(2, -2)$.

Choose three points on each graph. Explain how to use the three points to draw a vertical reflection of the graph.

7.

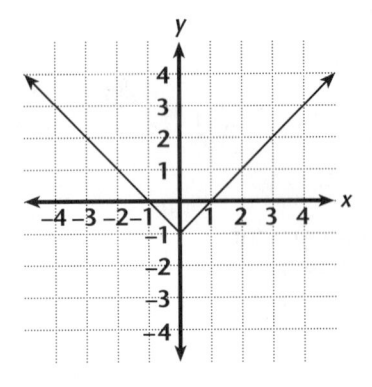

8.

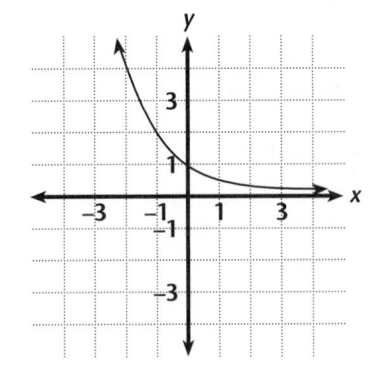

172 Reteaching Masters   Algebra 1

NAME _____ CLASS _____ DATE _____

Reteaching
14.5 Combining Transformations

◆ **Skill A** Identifying parent functions in transformations

Recall Transformations of a function are indicated by the addition or subtraction of constants from the variable term or from the entire function or by multiplication or division of the variable term by a constant.

◆ **Example**
In the following equations, identify the parent function:
a. $y = -3|x - 2| + 5$ **b.** $y = -(x + 3)^2 - 4$

◆ **Solution**
a. Identify the additions, multiplications, subtractions, or divisions that occur. If the addition of 5 is removed, the equation becomes $y = -3|x - 2|$. If the multiplication by -3 is removed, the equation becomes $y = |x - 2|$. Finally, if the subtraction of 2 is removed, the equation becomes $y = |x|$. This is the absolute-value parent function.

b. Start with $y = -(x + 3)^2 - 4$ and remove the additions and subtractions, starting with the subtraction of 4 outside the parentheses. This leaves $y = -(x + 3)^2$. Then remove the negative sign preceding the parentheses, leaving $y = (x + 3)^2$. Finally, remove the addition of 3 within the parentheses, producing the function $y = x^2$. This is the quadratic parent function.

Identify the parent function for each of the following:

1. $y = -2|x + 1| - 4$ _____

2. $y = 3(x - 1)^2 - 2$ _____

3. $y = 3 \cdot 2^{-x} + 1$ _____

4. $y = -3(x + 2) - 4$ _____

5. $y = \frac{3}{x} + 2$ _____

6. $y = \frac{3}{x + 2}$ _____

7. $y = 3x^2 - 4$ _____

8. $y = -2(x - 1)^2$ _____

Algebra 1 Reteaching Masters 173

NAME _____ CLASS _____ DATE _____

◆ **Skill B** Understanding the effect of order on combining transformations

Recall To determine the order of transformations to a function, reverse the order of operations. Addition or subtraction indicates a vertical translation; multiplication or division indicates a vertical stretch; addition or subtraction within parentheses or within absolute-value symbols indicates a horizontal translation.

◆ **Example**
Describe the various transformations included in the equation $y = 2|x - 1| + 3$.

◆ **Solution**
The first operation to consider is the addition of 3. This affects the parent function by translating it vertically 3 units up. The second operation, multiplication by 2, stretches the translated function by a factor of 2. The third operation, subtraction of 1, translates the stretched function horizontally 1 unit to the right. Thus, the parent function, $y = |x|$, has been shifted 1 unit to the right, stretched by a factor of 2, and then shifted 3 units up.

Describe the transformations of the parent functions included in each equation.

9. $y = -3|x + 2| - 3$ _____

_____

_____

10. $y = 2(x - 3)^2 + 1$ _____

_____

_____

11. $y = 4|x - 1| + 2$ _____

_____

_____

12. $y = 4 \cdot 2^x - 2$ _____

_____

_____

174 Reteaching Masters Algebra 1

ANSWERS

Reteaching—Chapter 1

Lesson 1.1

1. 30, 38
2. 45, 53
3. 8, 3
4. 8, 2
5. 48, 56
6. 85, 82
7. 65, 81
8. 36, 49
9. −48, −60
10. −28, −33
11. 49, 71
12. −21, −32
13. 12, 18, 27
14. 8, 11, 15, 20, 26

Lesson 1.2

1.

x	1	2	3	4	5
y	8	16	24	32	40

2.

x	1	2	3	4	5
y	11	17	23	29	35

3.

x	1	2	3	4	5
y	25	24	23	22	21

4.

x	1	2	3	4	5
y	120	60	40	30	24

5.

x	1	2	3	4	5
y	6	11	16	21	26

6.

x	1	2	3	4	5
y	−4	−9	−14	−19	−24

7. $x = 4$
8. $x = 9$
9. $x = 7$
10. $x = 12$
11. $x = 15$
12. $x = 18$
13. $4x = 18$; $4.5
14. $5 + 3x = 17$; 4 hours

Lesson 1.3

1. 14
2. 4
3. 21
4. 4
5. 12
6. 12

Algebra 1   Answers   **175**

ANSWERS

7. 41
8. 20
9. 13
10. 17
11. 26
12. 9
13. 26
14. 0
15. 10
16. 3
17. 19
18. 0
19. 2
20. 27
21. 23
22. 18
23. 20
24. 67
25. 96
26. 131
27. 49
28. 34
29. 16
30. 39

Lesson 1.4

1. $P(1, 4)$
2. $Q(-3, 3)$
3. $R(4, 0)$
4. $S(0, -4)$
5. $T(5, -4)$
6. $U(-4, -3)$
7. $V(-2, 0)$
8. $W(0, 3)$
9. yes
10. No; the line containing two of the points does not contain the third point.
11. yes
12. yes
13. Let d represent the number of pizza that Mark can buy and let t represent the total cost of the pizzas bought. $t = 5d$
14. Yes; if you substitute several different values for d and plot them on a grid, they will lie along a straight line.

Lesson 1.5

1. $2; y = 2x$
2. $-4; y = 25 - 4x$
3. $-3; y = -1 - 3x$
4. $5; y = -25 + 5x$
5.

1	2	3	4	5
-1	-2	-3	-4	-5

6.

1	2	3	4	5
3	1	-1	-3	-5

7.

1	2	3	4	5
15	20	25	30	35

ANSWERS

Lesson 1.6

1. positive
2. negative
3. none
4. The line should be close to all the points; check students' work.
5. The line should be close to all the points; check students' work.
6. Line b

Reteaching—Chapter 2

Lesson 2.1

1. $2.45 > -2.45$
2. $\frac{4}{5} < 1\frac{4}{5}$
3. $2.6 = 2\frac{3}{5}$
4. $-2 = -\frac{4}{2}$
5. $-11 > -12$
6. $2\frac{2}{7} > -\frac{16}{7}$
7. $0 < 4.5$
8. $0 > -4.5$
9. $-2.33; 2.33$
10. $-\frac{1}{17}; \frac{1}{17}$
11. $\frac{9}{2}; \frac{9}{2}$
12. $2\frac{6}{13}; 2\frac{6}{13}$
13. $-12.56; 12.56$
14. $12.56; 12.56$
15. $-1200; 1200$
16. $0.13; 0.13$
17. $1356; 1356$
18. $-3\frac{99}{100}; 3\frac{99}{100}$
19. $22.7; 22.7$
20. $-100\frac{1}{2}; 100\frac{1}{2}$
21. 2.8
22. -11
23. 30
24. -4
25. 0
26. 6.25
27. 5
28. 25
29. 4

Lesson 2.2

1. 3
2. -8
3. 1
4. -6
5. -7
6. -1
7. -19
8. -10
9. -42
10. -25
11. -11
12. 6
13. -22

Algebra 1   Answers   **177**

ANSWERS

14. −9.5
15. 3.1
16. 8.6
17. $\frac{1}{3}$
18. $-\frac{4}{5}$
19. $-\frac{5}{8}$

Lesson 2.3

1. 10
2. −6
3. −10
4. 3
5. −2
6. −24
7. 43
8. 40
9. 48
10. −2
11. −90
12. −6.1
13. −2.8
14. −1.4
15. 1.7
16. 12.5
17. −15.8
18. 10
19. 5

20. 15
21. 20
22. 55
23. 12
24. 1.7
25. 2.7
26. $2\frac{3}{4}$

Lesson 2.4

1. 14
2. −88
3. −84
4. 39
5. 300
6. −750
7. −60
8. 0
9. 11
10. 3
11. −0.38
12. −14.4
13. 3
14. −5
15. −4
16. 12
17. 125
18. −3

ANSWERS

19. 32

20. 0

21. 0.15

22. 27

23. −2

24. $-\frac{8}{9}$

Lesson 2.5

1. Associative Property of Addition
2. Distributive Property
3. Commutative Property of Addition
4. Associative Property of Multiplication
5. Commutative Property of Addition
6. 12
7. 10
8. 9
9. 1.9
10. 1.8
11. 5
12. 15
13. 4.3
14. 148
15. 480
16. 1300
17. 0
18. 56
19. 86

20. 170
21. 120
22. 170
23. 130

Lesson 2.6

1. $4x + 9$
2. $7m - 2$
3. $22 - 3y$
4. $5b - 4a$
5. $-6x - 22$
6. $7n + 6m + 5$
7. $3x + 8$
8. $2m - 2$
9. $-4p + 12q - 14r$
10. $26x - 6y$
11. $-5p$
12. $-y - 17$
13. $-3x - 4$
14. $-4m - 4n$
15. $9x - 19y$
16. $-12a + 17b$
17. $-3p + 2n + t$
18. $6x + y + 4$
19. $10a - 10b - 10$
20. $5m + 5n - 12$

Algebra 1

ANSWERS

Lesson 2.7

1. $-22x^2$
2. $14x^2$
3. $-2x^2 - 2x$
4. $-18x^2 + 12x$
5. $x^2 + 4x$
6. $3x^2 - 24x$
7. $10x^2 + 50x$
8. $7x^2 - 35x$
9. $-2x^2 - 2x$
10. $x^2 - 8x$
11. $-7a$
12. $-45n$
13. $-3x - 2$
14. $-2k + 3$
15. $-14x + 6y$
16. $5x + 30y$
17. $x + 2y$
18. $9x - 24$
19. $3 + 6x$
20. $3x - 9y$
21. $8.5y + 3.6$
22. $-20a - 30$

Reteaching—Chapter 3

Lesson 3.1

1. $t = 56$
2. $x = -44$
3. $y = 6.1$
4. $m = -40$
5. $r = -16$
6. $x = -5$
7. $z = \frac{2}{5}$
8. $a = -3$
9. $b = -11.5$
10. $x = \frac{1}{10}$
11. $x + 2790 = 5000$; $2210
12. $35 + 90 + c = 180$; $m\angle C = 55$
13. $x = 18$
14. $t = -26$
15. $y = 25$
16. $m = 20.4$
17. $b = -9.5$
18. $x = 640$
19. $y = 8\frac{1}{2}$
20. $a = -5$
21. $x = 1.6$
22. $y = 7$
23. $b = 180$
24. $c = -10.7$
25. $x - 60 = 215$, $275

ANSWERS

Lesson 3.2

1. $d = 7$
2. $y = -5$
3. $a = -7$
4. $x = -12$
5. $s = 17$
6. $t = 21\frac{1}{3}$
7. $x = 12$
8. $x = -4\frac{2}{5}$
9. $3\frac{1}{2}$ inches, or 3.5 inches
10. $22\frac{1}{2}$ hours
11. $m = 55$
12. $y = 70$
13. $a = 64$
14. $x = -68$
15. $t = -312$
16. $n = -48$
17. $x = -50.4$
18. $y = 48.02$
19. -49.2
20. 45

Lesson 3.3

1. $x = 2$
2. $t = 6$
3. $y = 4$
4. $x = 5$
5. $x = 5$
6. $z = 6$
7. $s = 16$
8. $w = 9$
9. $f = -9$
10. $c = 7$
11. $x = 6$
12. $t = -81$
13. $x = 0.5$
14. $x = 4$
15. $j = 5$
16. $k = -5$
17. $80 + 6s = 290$; each scarf cost $35.
18. $35 + 8c = 155$; each train car cost $15.
19. $0.50x + 100 = 150$; they would need to sell 100 streamers.

Lesson 3.4

1. $x = 1$
2. $y = -8$
3. $a = -4$
4. $x = 2$
5. $t = -9$
6. $x = 1$
7. $y = 2.5$
8. $b = -1$
9. $w = 2$
10. $p = 2$

Algebra 1   Answers   **181**

ANSWERS

11. $x = 4$
12. $x = \frac{1}{8}$
13. $n = -\frac{7}{15}$
14. $m = 6$
15. $x = -16$
16. $y = 36$
17. She must score 93.

Lesson 3.5

1. $x = -6$
2. $t = 5$
3. $z = 3$
4. $k = -13.5$
5. $x = 4$
6. $m = 2$
7. $h = 3$
8. $n = -3$
9. $t = 3$
10. $c = -5$
11. $f = -1$
12. $y = 7$
13. $3(x - 1.50) = 35.97$; the original cost is $13.49.
14. $12x = 2(x + 6)$; $x = 1.2$
15. $4(x - 2) = 20$; the original average cost was $7.
16. $29 + 0.15x = 20 + 0.25x$; the rates are equal when you drive 90 miles in one day.

Lesson 3.6

1. $x = y + 10$
2. $y = z - x$
3. $z = x + y$
4. $x = 2y - a$
5. $y = 32 - x + z$
6. $t = \frac{d}{r}$
7. $h = \frac{2A}{b}$
8. $x = \frac{5}{3}y + 5$
9. $y = 2x - 3$
10. $l = \frac{P}{2} - w$
11. $w = 3$
12. $r = 7$
13. 10 meters
14. 4 hours
15. $7000
16. 16 kPa

Reteaching—Chapter 4

Lesson 4.1

1. $\frac{2}{5}$
2. $\frac{5}{1}$
3. $\frac{9}{8}$
4. $\frac{3}{2}$
5. $\frac{6}{17}$

ANSWERS

6. $\frac{2}{15}$

7. $\frac{4}{5}$

8. $\frac{2}{5}$

9. $\frac{18}{1}$

10. $\frac{1}{5}$

11. $\frac{2}{61}$

12. $\frac{3}{500}$

13. no; cross products: 28 and 35

14. yes; cross products: 72 and 72

15. yes; cross products: 120 and 120

16. no; cross products: 400 and 500

17. no; cross products: 336 and 352

18. no; cross products: 540 and 450

19. yes; cross products: 528 and 528

20. yes; cross products: 540 and 540

21. yes; cross products: 2700 and 2700

22. $b = 6.3$

23. $a = 1$

24. $p = 50$

25. $r = 24$

26. $k = 60$

27. $t = 2.1$

28. $d = 12$

29. $m = 15$

30. $s = 9$

31. $e = 10$

32. $n = 3$

33. $z = 270$

34. 18 free throws

Lesson 4.2

1. 0.86

2. 0.783

3. 0.06

4. 0.46

5. 0.23

6. 0.9

7. 1.25

8. 0.004

9. $\frac{4}{5}$

10. $\frac{1}{8}$

11. $\frac{9}{500}$

12. $\frac{1}{100}$

13. $\frac{12}{25}$

14. $\frac{2}{5}$

15. $1\frac{4}{5}$

16. $\frac{3}{8}$

17. $\frac{7}{20}$

18. $\frac{1}{3}$ off

19. 32

20. 40

Algebra 1   Answers   183

ANSWERS

21. 25.6%
22. 40%
23. 90
24. 150%
25. 127.5
26. 62.5%
27. $32.25
28. $70.80

Lesson 4.3

1. 24%
2. 32%
3. 16%
4. 8%
5. 40%
6. 80%
7. about 56%
8. about 44%
9. 32%
10. about 11%
11. 16%
12. about 9%
13. 24%
14. about 35%
15. 40%
16. 60%
17. 30%
18. 10%
19. 60%
20. 40%
21. 60%
22. 40%

Lesson 4.4

1. mean = 908; median = 900; mode = 900; range = 334
2. mean = 22.75; median = 22; mode = 22; range = 12
3. mean = 13.75; median = 14; mode = 14; range = 28
4. mean = 8; median = 8; mode = 5 and 10; range = 6
5. Frequency Table:

Scores	70	75	80	89	91	95	100
Frequency	II	IIII	II	III	I	II	I

   mean = 83.2; median = 80; mode = 75; range = 30

6. Frequency Table:

Accidents	0	1	2	3	4	5	6
Frequency	I	II	III	III	I		II

   mean = 2.75; median = 2.5; mode = 2 and 3; range = 6

Lesson 4.5

1. It could be misleading because the science test is a 200-point test and the mathematics test is a 100-point test. Thus, the scales are different.
2. 200 points; 1 person scored 200 points.
3. 80 points
4. 180 points

ANSWERS

5. 180 people

6. 150 people

7. 13 people

8. rock

Lesson 4.6

1. 30

2. 84

3. 82.9

4. 80

Stems	Leaves
6	5
7	0 0 2
8	0 0 0 2 6 6
9	0 0 2 4 5 5

5. 0; 10; 7; 3; 8

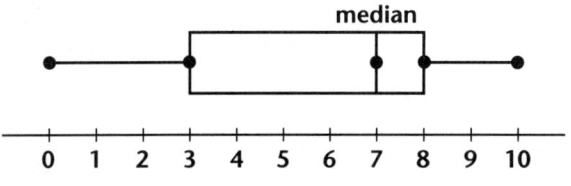

6. 76; 98; 85.5; 83; 93

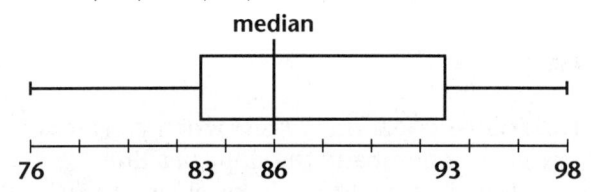

Reteaching—Chapter 5

Lesson 5.1

1. function
domain: 1.3, 3.1, 10, 12
range: −1.3, 2.3, 10, 21

2. function
domain: 1, 2, 3, 4, 5, 6
range: 64, 32, 16, 8, 4, 2

3. not a function
domain: −3, −5, 8, 10
range: 7, 8

4. function
domain: 18, 20, −20, −7, 8
range: 11, 12, −8, −1.5

5. $y = 32$ and $x = \frac{3}{2}$

6. $y = -30$ and $x = -\frac{63}{2}$

7. $y = 6x - 1$

8. $y = 4x + 0.4$

Lesson 5.2

1. negative

2. positive

3. undefined

4. 0

5. $\frac{1}{2}$

6. −2

7. 1

8. −1

9. 2

10. $\frac{1}{2}$

11. $\frac{3}{5}$

12. $-\frac{4}{7}$

Lesson 5.3

1. $\frac{4}{3}$

2. $\frac{1}{3}$

3. 2

Algebra 1   Answers   **185**

ANSWERS

4. $\frac{4}{5}$
5. 0
6. $\frac{9}{5}$
7. $\frac{3}{2}$
8. undefined
9. $\frac{4}{5}$
10. 6.5; $y = 6.5x$; $78, $97.50
11. $\frac{1}{7}$; $y = \frac{1}{7}x$; 21, 20

Lesson 5.4

1. slope = 3; y-intercept = -1
2. slope = $\frac{1}{2}$; y-intercept = 2
3. slope = -1; y-intercept = $\frac{1}{2}$
4. $y = 2x - 1$
5. $y = \frac{1}{3}x - 3$
6. $y = 2x - 1$
7. $y = -3x + 2$
8. $y = \frac{1}{2}x + 3$
9. $y = x - 1$
10. $y = 3x + 1$
11. $y = -\frac{2}{3}x - 2$
12. $y = -\frac{2}{3}x - 2$
13. $y = 3x + 5$
14. $20
15. 7

Lesson 5.5

1. $5x - 10y = 15$
2. $2x - 3y = 0$
3. $2x - 5y = 5$
4. $6x - y = 12$
5. $(0, -2); (-1, 0)$
6. $(0, -3); (2, 0)$

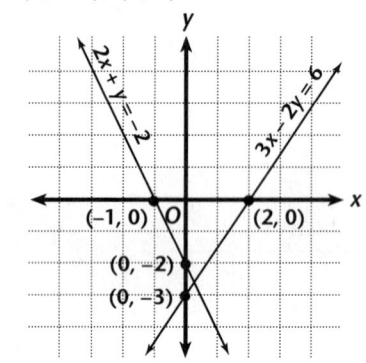

7. $y + 1 = \frac{1}{2}(x - 4)$
8. $y - 6 = 2x$ or $y = 2(x + 3)$
9. $y - 2 = \frac{1}{3}(x - 3)$ or $y + 1 = \frac{1}{3}(x + 6)$
10. $y = \frac{1}{2}x - 3$
11. $y = -\frac{1}{2}x + 6$
12. $y = \frac{1}{3}x + 1$
13. Point-slope form is easier when you are given a point and the slope of a line. Slope-intercept form is easier when you are given the slope and the y-intercept of a line.

Lesson 5.6

1. $-\frac{1}{2}$
2. -3
3. $\frac{2}{3}$
4. -4

186 Answers Algebra 1

ANSWERS

5. $y + 4 = 3(x + 1)$
6. $y + 4 = \frac{1}{2}(x - 2)$
7. $x = -2$
8. $y - 15 = \frac{3}{2}(x - 4)$
9. $y = -6$
10. $-\frac{1}{2}$
11. -3
12. $\frac{4}{3}$
13. 2
14. undefined
15. 0
16. $y - 3 = \frac{1}{2}(x - 2)$
17. $x = 1$
18. $y - 4 = -\frac{3}{2}(x - 3)$
19. $y - 1 = 2(x - 4)$
20. $y = 6$

Reteaching—Chapter 6

Lesson 6.1

1. $x > 60$
2. $m \geq 25.5$
3. $m \geq 1$
4. $a < -8$
5. $z \leq 4$
6. $b \leq -14$
7. $x > 30$
8. $b \leq 13$
9. $y > 15$
10. $s > 67.1$
11. $z \leq -5$
12. $x > 61,450$
13. $200 \leq S \leq 600$
14. $6 \leq S < 30$
15. $45 \leq S \leq 99$
16. $9 < S < 23$

Lesson 6.2

1. $y \leq 15$
2. $a > 8$
3. $m < 8$
4. $b \leq -3$
5. $r > -12$
6. $d \geq -40$
7. $x < 8$
8. $p \geq -5$
9. $y \leq 3$
10. $x + 5 \geq 8; x \geq 3$
11. $0.12x \geq 5; 50$
12. $9x \leq 85$; no more than 9 T-shirts
13. $1.65c \leq 20$; no more than 12 cards
14. $4.75x \geq 120$; at least 26 hours
15. $y \leq 5$
16. $a > 8$
17. $m \leq 5$

Algebra 1   Answers **187**

ANSWERS

18. $b < 60$

19. $x \geq 10$

20. $p < 7$

21. $z \geq -5$

22. $d < 2$

23. $m \leq 4$

24. $x < 2$

25. Let x represent the range of numbers.
    $3x + 8 < -7; x < -5$

26. Let x represent the number of gallons of juice that Sam can buy.
    $2.50x + 8 \leq 20$; 4 gallons or less

27. Let x represent the number of T-shirts that Sue can buy.
    $9x + 30 \leq 50$; 2 T-shirts or less

28. Let x represent the number of bottles of soda that Rafael and Sally need to bring.
    $2x + 7 \geq 25$; at least 9 bottles each

Lesson 6.3

1. 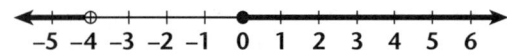
2. 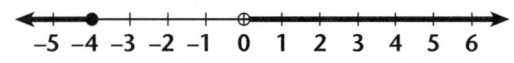
3. 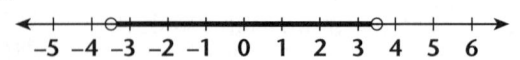
4. 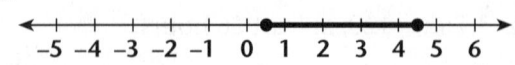
5.

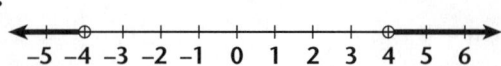

6. 
7. 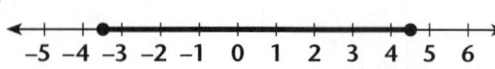
8. 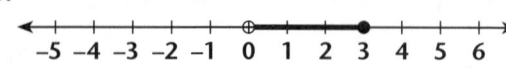
9. $1 \leq x < 1.5$
10. $1 < x < 2$
11. $x < 0$ or $x \geq 3.5$
12. all real numbers

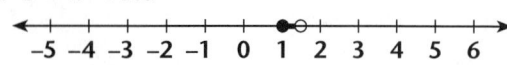

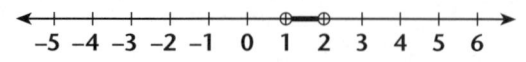

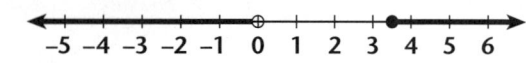

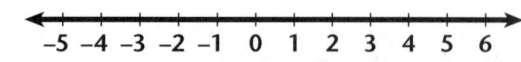

Lesson 6.4

1. 8
2. 15
3. 7
4. 7
5. 9
6. 9
7. 16
8. 16
9. 0
10. 19
11. 19
12. 10

ANSWERS

13. 40

14. 10

15. Domain: all real numbers
Range: all nonnegative numbers

16. Domain: all real numbers
Range: all nonnegative numbers

17. Domain: all real numbers
Range: all nonpositive numbers

18. Domain: all real numbers
Range: $y \geq -3$

19. Domain: all real numbers
Range: $y \leq 1$

20. Domain: all real numbers
Range: all nonnegative numbers

21. Domain: all real numbers
Range: $y \geq 6$

22. Domain: all real numbers
Range: all nonnegative numbers

Lesson 6.5

1. $x = 8$ or $x = -12$

2. $x = 4$ or $x = 14$

3. $x = 1$ or $x = 5$

4. $x = 15$ or $x = 9$

5. $x = 6$ or $x = 4$

6. $x = 11$ or $x = -25$

7. $x = 6$ or $x = -5$

8. $x = 3$ or $x = 2\frac{1}{3}$

9. $x = 4$ or $x = -5$

10. $x = 2$ or $x = -18$

11. $x = 4$ or $x = -\frac{4}{5}$

12. $x = -3$ or $x = 2\frac{1}{2}$

13. $x < -7$ or $x > 9$

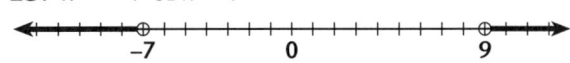

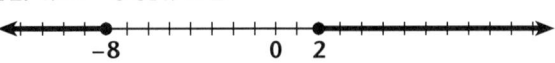

14. $x \leq -8$ or $x \geq 2$

15. $4 < x < 8$

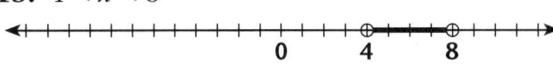

16. $-6 \leq x \leq -4$

17. $-1 < x < 5$

18. $x < 4$ or $x > 10$

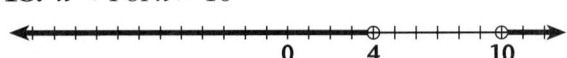

Reteaching—Chapter 7

Lesson 7.1

1. (2, 2)

2. (1, 2)

3. about 76 feet

4. about 7.3 meters by 12.7 meters

Lesson 7.2

1. $(-1, -4)$

2. (4.5, 2)

3. (12, 1)

4. (2, 1)

5. (3, 5)

6. $(-9, -7)$

Algebra 1   Answers   **189**

ANSWERS

7. $\begin{cases} x + y = 346 \\ x = \frac{1}{3}(y - 6) \end{cases}$
   85 and 261

8. $\begin{cases} x + y = 90 \\ x = \frac{2}{3}y \end{cases}$
   54° and 36°

9. $\begin{cases} r = 3s \\ r + s = 28 \end{cases}$
   Raul is 21 years old and Sara is 7 years old.

10. $\begin{cases} x + y = 8000 \\ 0.07x + 0.05y = 500 \end{cases}$
    $5000 at 7% and $3000 at 5%

Lesson 7.3

1. (10, 15)
2. (−1, 2)
3. (0.2, 8)
4. (4, −0.5)
5. (2, −1)
6. (4, 3)
7. (5, 1)
8. (2, 0)
9. (1.2, −0.6)
10. (2, 2)
11. (6, 4)
12. (4, −1)
13. $\begin{cases} t + c = 114 \\ 12t + 9c = 1242 \end{cases}$
    72 T-shirts and 42 baseball caps

Lesson 7.4

1. inconsistent
2. (3, 1); consistent
3. (0, 2); consistent
4. inconsistent
5. consistent and dependent
6. inconsistent
7. (−6, 8); consistent and independent
8. inconsistent
9. consistent and dependent
10. consistent and dependent
11. (3, 4); consistent and independent
12. (0.5, 1.2); consistent and independent

Lesson 7.5

1.

2.

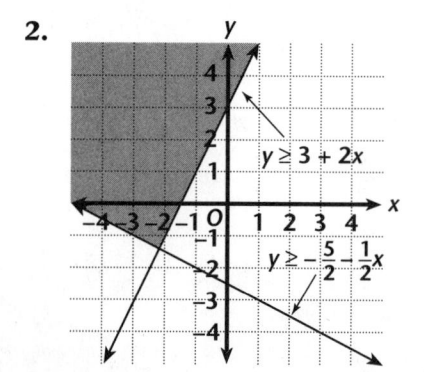

ANSWERS

3.

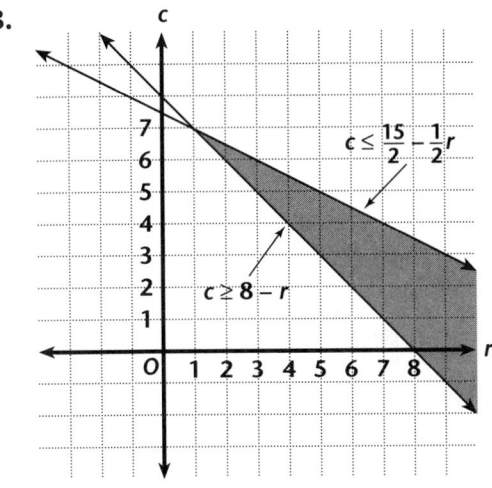

Answers may vary. Sample answer: (5, 5) and (8, 2)

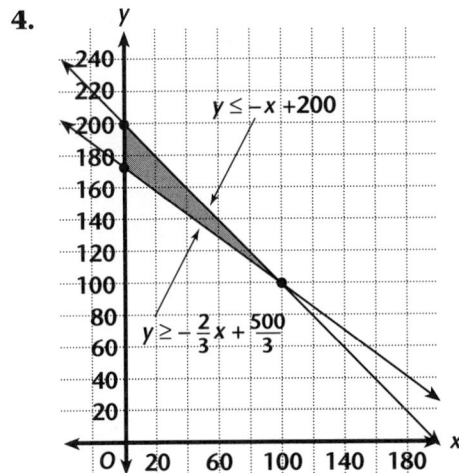

4.

Answers may vary. Sample answer: (180, 20) and (100, 100)

Lesson 7.6

1. $\begin{cases} x = y + 1 \\ x + y = 27 \end{cases}$
   Joel is 13 years old; Roberto is 14 years old

2. $\begin{cases} x = 3y \\ x + 6 = 2(y + 6) \end{cases}$
   Latisha is 18 years old now.

3. $\begin{cases} p + n = 260 \\ p + 5n = 432 \end{cases}$
   43 nickels and 217 pennies

4. $\begin{cases} n + q = 57 \\ 5n + 25q = 725 \end{cases}$
   22 quarters and 35 nickels

5. $\begin{cases} x + y = 42 \\ 32x + 40y = 1440 \end{cases}$
   thirty 32-cent stamps and twelve 40-cent stamps

6. $\begin{cases} t + u = 12 \\ u = 3t \end{cases}$
   39

7. $\begin{cases} t + u = 10 \\ 2(10t + u) - 1 = 10u + t \end{cases}$
   37

Reteaching—Chapter 8

Lesson 8.1

1. 625
2. 81
3. 1,000,000
4. 128
5. 32,768
6. 729
7. 6
8. 10,000
9. 144
10. 1
11. $3^9 = 19,683$
12. $2^7 = 128$
13. $10^8 = 100,000,000$
14. $5^7 = 78,125$
15. $8^{10} = 1,073,741,824$
16. $4^6 = 4096$
17. $15a^5$
18. $-21c^3d^2$

Algebra 1   Answers   **191**

ANSWERS

19. $5s^3t^5$
20. $24p^7q^3$
21. $4m^5n^4$
22. $6a^6b^5c^2$

Lesson 8.2

1. 4096
2. 729
3. y^{12}
4. m^{10}
5. $4s^6$
6. $5r^{10}$
7. $27c^{18}$
8. n^{3d}
9. 1,000,000
10. $32y^{15}$
11. $216x^{12}$
12. $512q^9$
13. c^8d^8
14. $81m^2n^{10}$
15. $4e^{12}f^3$
16. $16p^{20}r^{12}$
17. $9y^6$
18. $-g^5h^{20}$
19. $-a^4b^9$
20. $-432c^{11}d^{18}$

Lesson 8.3

1. 343
2. 32
3. $4d^6$
4. $\dfrac{m^2}{2}$
5. 1,000,000,000
6. $-9r^4$
7. $6c^7$
8. $-9g^7$
9. xy^2
10. p^4qr^2
11. $-6g^6h^2$
12. $-\dfrac{3y^2z^4}{4}$
13. $3s^4t^3$
14. $4.2b^3c^5$
15. $\dfrac{16r^{12}}{n^4}$
16. $-343c^6m^3$
17. $-125c^3$
18. $d^{10}e^{12}$

Lesson 8.4

1. 1
2. $\dfrac{1}{25}$
3. 1
4. $\dfrac{1}{4}$
5. $\dfrac{1}{27}$
6. 1

ANSWERS

7. $\frac{1}{125}$

8. $\frac{1}{64}$

9. a^{-2}

10. c^{-5}

11. y^{-2}

12. m^{-9}

13. p^8

14. q^{-5}

15. x^{-11}

16. z^3

17. t^5

18. 3125

19. x^{-5}

20. $3^{-1} = \frac{1}{3}$

21. t^{-3}

22. 1,048,576

23. 25

24. a^{-3}

25. r^3

26. $\frac{1}{1024}$

Lesson 8.5

1. 4.57×10^9

2. 2.3×10^{-6}

3. 4.58×10^{-3}

4. 6.2×10^7

5. 7.05×10^{10}

6. 8.75×10^{-5}

7. 5.8×10^3

8. 2.6×10^{-2}

9. 3.5×10^7

10. 7.2×10^{-8}

11. 2.07×10^{12}

12. 3.05×10^{-3}

13. 2.4×10^8

14. 5×10^2

15. 2.236×10^7

16. 4×10^4

17. 2.3646 E 08

18. 3 E 02

19. 2.686 E 12

20. 2 E 02

Lesson 8.6

1. about 793,958 people

2. about 900,526 people

3. about 1,246,182 people

4. about 428,765 people

5. about 12,628,582 people

ANSWERS

6. The function increases as x increases.

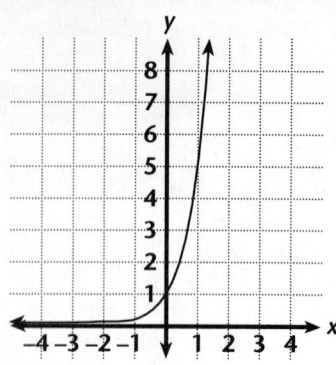

7. The function decreases as x increases.

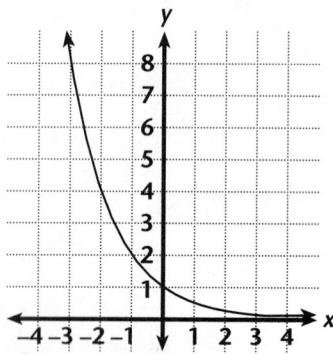

8. The function increases as x increases.

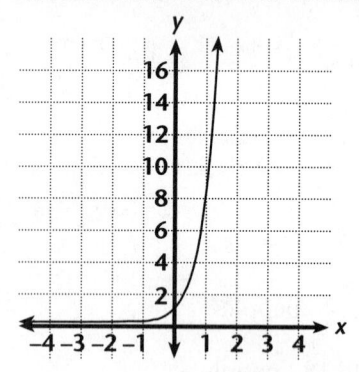

9. The function decreases as x increases.

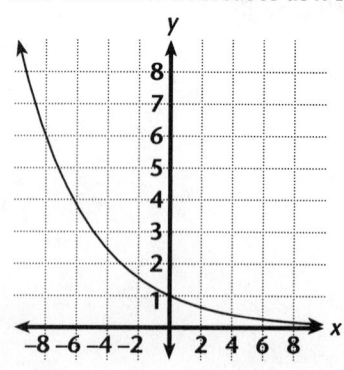

Lesson 8.7

1. $P = 24(1 - 0.06)^5$, where P represents present value

2. $P = 500,000(1 + 0.02)^{-6}$, where P represents present population

3. $P = 54(1 + 0.04)^7$, where P represents present value

4. $9030.56

5. about 29,364 people

6. 6480 people

Reteaching—Chapter 9

Lesson 9.1

1. $6b^2 + b$

2. $10c^2 + c$

3. $5b^2 - 3b^2$

4. $5y^3 + 5y^2 + 6y - 1$

5. $2r^3 + 6r^2 + 7r + 6$

6. $7m^3 - 5m^2 - 4m - 5$

7. $-3x^2 + 4x - 3$

8. $4x^2 - 2x + 9$

9. $3x - 1$

10. $7x^3 + 1$

11. $-3c^2 - c - 5$

12. $-n^2 + 2n - 3$

13. $2z^2 + z + 1$

14. $-5r^2 - 4r + 9$

15. $-4t^2 + t$

16. $9q^2 + q + 3$

ANSWERS

17. $-5 + 2a + 3a^2$
18. $-5e^3 + 4e^2 - 2e$
19. $-2x^2 + 2x + 8$
20. $-8x + 3$
21. $6x^2 - x + 3$
22. $x^2 + 3x - 7$
23. $3x^2 + 2$
24. $-x + 6$

Lesson 9.2

1. $2x + 6$
2. $6x - 6$
3. $x^2 - 3x$
4. $-x^2 + x$
5. $2x^2 + 4x$
6. $3x^2 - 3x$
7. $2x^2 - 4x$
8. $6x^2 + 4x$
9. $x^2 + 3x + 2$
10. $x^2 + x - 2$
11. $x^2 - x - 2$
12. $x^2 - 3x + 2$
13. $x^2 + 6x + 9$
14. $x^2 - 6x + 9$
15. $x^2 - 9$
16. $x^2 - 9$
17. $2x^2 + 5x + 3$
18. $2x^2 - 5x + 2$
19. $3x^2 - 4x - 4$
20. $6x^2 - x - 2$

Lesson 9.3

1. $4x + 20$
2. $5x - 10$
3. $2x^2 - 2x$
4. $6x^2 + 2x$
5. $-5x^2 + 30x$
6. $3x^2 + 9x$
7. $x^2 + 5x + 4$
8. $x^2 + 5x + 6$
9. $x^2 + 2x - 15$
10. $2x^2 + 7x + 6$
11. $3x^2 - 18x + 15$
12. $12x^2 - 25x + 12$
13. $x^2 + 7x + 10$
14. $x^2 - x - 12$
15. $x^2 - 8x + 15$
16. $2x^2 + 3x - 9$
17. $4x^2 - 20x + 25$
18. $25x^2 - 40x + 16$
19. $4x^2 - 3x - 1$
20. $2x^2 - 7x + 6$
21. $-6x^2 + 17x - 12$

Algebra 1

ANSWERS

Lesson 9.4

1. 0
2. 54
3. 12
4. 1
5. 75 cubic meters
6. 110 square meters
7. 125 cubic feet
8. 212.5 square feet
9. 0, 0, 2, 6, 12
10. 0, −4, −6, −6, −4
11. 28, 18, 10, 4, 0
12. 4, 9, 16, 25, 36
13. 0, −3, −4, −3, 0
14. 9, 4, 1, 0, 1

Lesson 9.5

1. 1, 2, 3, 4, 6, 12; no
2. 1, 5, 7, 35; no
3. 1, 47; yes
4. 1, 3, 19, 57; no
5. 1, 7, 11, 77; no
6. 1, 97; yes
7. $3m(m - 7)$
8. $t(8t + 15)$
9. $3(6p^2 + 7p + 3)$
10. $4d(d^2 - 5d + 2)$
11. $(x + 8)(x + 5)$
12. $(x - 7)(x - 3)$
13. $(x + 3)(x + 4)$
14. $(x - 2)(x - 3)$
15. $(x + 5)(x - 2)$
16. $(x - 4)(x + 1)$

Lesson 9.6

1. $x^2 + 8x + 16$
2. $4v^2 - 20v + 25$
3. $9d^2 - 6ad + a^2$
4. $100k^2 - 20kt + t^2$
5. $(x + 1)^2$
6. prime
7. $(2p + 4)^2$
8. $(2a - b)^2$
9. $(r - 2s)^2$
10. $(7b - 3c)^2$
11. $t^2 - 25$
12. $m^2 - d^2$
13. $9r^2 - s^4$
14. $25p^2 - 16q^2$
15. $(t + 7)(t - 7)$
16. $(2 - a)(2 + a)$
17. prime
18. $(2s - t)(2s + t)$
19. $(5c - 2q)(5c + 2q)$
20. $(4b + c^2)(4b - c^2)$

ANSWERS

21. $(mn + p)(mn - p)$

22. $(s^2 + t^2)(s + t)(s - t)$

Lesson 9.7

1. $(x + 1)(x + 2)$

2. $(x - 3)(x + 4)$

3. $(x - 7)(x - 3)$

4. $(x + 5)(x - 1)$

5. $(x + 8)(x + 3)$

6. $(x - 4)(x - 4)$

7. $(x - 2)(x + 1)$

8. $(x + 4)(x - 1)$

9. $(x + 3)(x + 1)$

10. $(x - 3)(x - 1)$

11. $(x + 4)(x - 2)$

12. $(x + 5)(x - 3)$

13. $(x + 5)(x - 3)$

14. prime

15. $(x + 3)(x - 4)$

16. $(x + 4)(x + 2)$

17. $(x - 2)(x - 18)$

18. $(x + 6)(x - 4)$

Lesson 9.8

1. 3 and 2

2. −5 and 4

3. −4 and 4

4. 6

5. 2.8 and −5.2

6. $-\frac{2}{3}$ and 1

7. 2

8. $-\frac{3}{5}$ and $-\frac{7}{4}$

9. −10 and $\frac{8}{7}$

10. −3 and 1.5

11. −2 and 6

12. 3

13. 2 and 7

14. −5 and −1

15. 2 and 5

16. −6 and 6

17. −4

18. −3 and 4

19. $-\frac{1}{3}$ and $\frac{1}{3}$

20. $-\frac{1}{2}$

Reteaching—Chapter 10

Lesson 10.1

1. vertex: (5, 1); axis of symmetry: $x = 5$

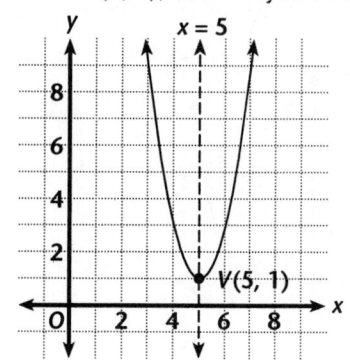

Algebra 1   Answers   **197**

ANSWERS

2. vertex: (4, −2); axis of symmetry: $x = 4$

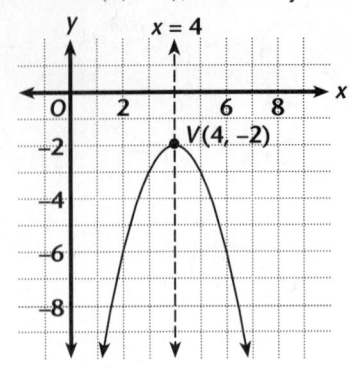

3. vertex: (−1, 4); axis of symmetry: $x = −1$

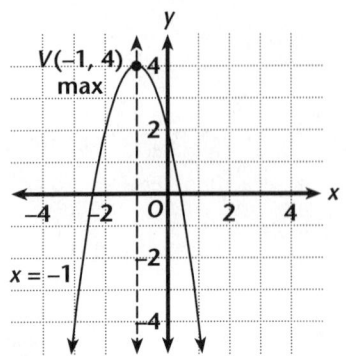

4. vertex: (3, 5); axis of symmetry: $x = 3$

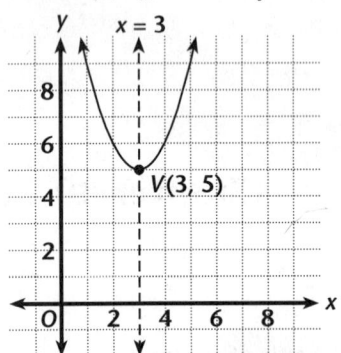

5. 1 and 2

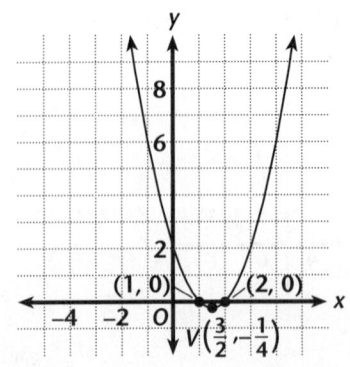

6. −2 and 5

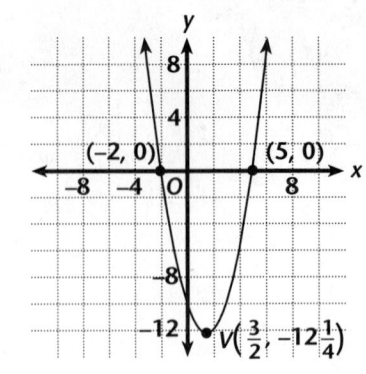

7. −3 and −1

8. none

9. $\frac{1}{2}$ and 2

10. −2 and $\frac{3}{2}$

Lesson 10.2

1. 6

2. −14

3. −16.58

4. 22

5. 15.49

6. −29

7. −5 and 5

8. −8.66 and 8.66

9. −10.39 and 10.39

10. −7 and 5

11. −6 and 14

12. −11 and 5

13. −16 and 2

14. −3.92 and 7.92

15. −10.18 and 12.18

198 Answers Algebra 1

ANSWERS

Lesson 10.3

1. $x^2 + 12x + 36$; -36
2. $x^2 + 5x + \frac{25}{4}$; $-\frac{25}{4}$
3. $x^2 - 20x + 100$; -100
4. $x^2 - 2x + 1$; -1
5. $x^2 - 10x + 25$; -25
6. $x^2 + 11x + \frac{121}{4}$; $-\frac{121}{4}$
7. $y = (x + 1)^2 + 0$; $V(-1, 0)$
8. $y = (x - 4)^2 - 13$; $V(4, -13)$
9. $y = (x + 2)^2 - 7$; $V(-2, -7)$
10. $y = (x - 1)^2 + 3$; $V(1, 3)$
11. $y = (x - 6)^2 - 72$; $V(6, -72)$
12. $y = (x + 1)^2 - 5$; $V(-1, -5)$
13. $y = \left(x - \frac{3}{2}\right)^2 + \frac{15}{4}$; $V\left(\frac{3}{2}, \frac{15}{4}\right)$
14. $y = \left(x - \frac{5}{2}\right)^2 - \frac{125}{4}$; $V\left(\frac{5}{2}, -\frac{125}{4}\right)$
15. $y = \left(x + \frac{1}{2}\right)^2 + \frac{3}{4}$; $V\left(-\frac{1}{2}, \frac{3}{4}\right)$

Lesson 10.4

1. -2 and 1
2. -5 and 2
3. -2 and -3
4. -1
5. $5 + \sqrt{22}$ and $5 - \sqrt{22}$
6. $-2 + \sqrt{7}$ and $-2 - \sqrt{7}$
7. -4 and 5
8. 3 and 4
9. -5 and 1
10. -2 and -4
11. -6 and 0
12. $-3 + \sqrt{11}$ and $-3 - \sqrt{11}$

Lesson 10.5

1. 1 and 4
2. -4 and 6
3. -3
4. -5 and 2
5. $-\frac{3}{2}$ and 2
6. $\frac{1 + \sqrt{33}}{4}$ and $\frac{1 - \sqrt{33}}{4}$
7. -4 and 2
8. $-\frac{5}{2}$ and 3
9. $\frac{1}{2}$ and $\frac{3}{2}$
10. 0; one solution
11. 9; two solutions
12. -7; no real solutions

Lesson 10.6

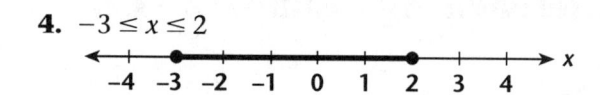

1. $-2 < x < -1$
2. $x \leq 3$ or $x \geq 5$
3. $x < -3$ or $x > 3$
4. $-3 \leq x \leq 2$

ANSWERS

5. 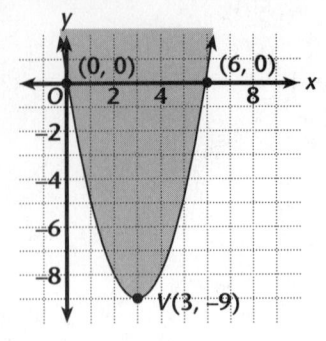 [graph showing parabola with points (0,0), (6,0), vertex V(3,−9)]

6. [graph showing parabola with points (−4,4), (0,4), vertex V(−2,0)]

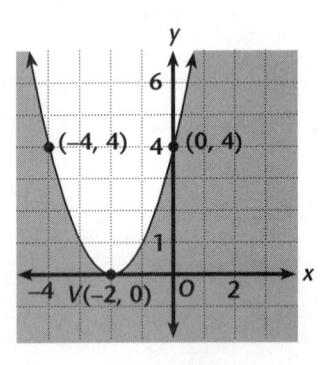

7. 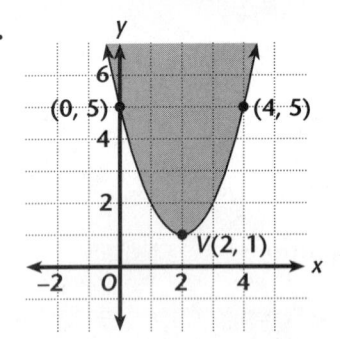 [graph showing parabola with points (0,5), (4,5), vertex V(2,1)]

8. 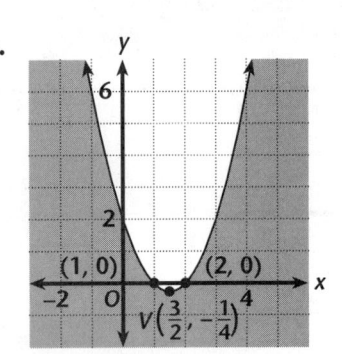 [graph showing parabola with points (1,0), (2,0), vertex V(3/2, −1/4)]

Reteaching—Chapter 11

Lesson 11.1

1. yes
2. yes
3. no
4. $x = 8$
5. $x = 10$
6. $x = 5$
7. 24 inches
8. 45 amps
9. $10\frac{2}{3}$ days
10. 25 meters

Lesson 11.2

1. -1
2. -6 and 1
3. -3 and 3
4. $-\frac{17}{3}$ and $\frac{1}{5}$
5. $\frac{1}{6}$ and $\frac{1}{2}$
6. $\frac{1}{6}$ and undefined
7. shifted down 3 units
8. shifted to the right 1 unit
9. shifted up 2 units and to the left 1 unit
10. stretched by a factor of 2 and shifted to the left 2 units
11. stretched by a factor of 3, shifted to the left 5 units, and shifted up 2 units
12. shifted down 5 units and to the right 3 units

Lesson 11.3

1. $\frac{2t}{t-1}$, $t \neq 1$
2. $\frac{m+3}{2(m-2)}$, $m \neq 2$

200   Answers   Algebra 1

ANSWERS

3. $\frac{m}{2+3m}$, $m \neq 0$, $m \neq -\frac{2}{3}$

4. $\frac{3}{y+2}$, $y \neq -2$

5. $\frac{3}{x+1}$, $x \neq -1$

6. $\frac{4}{r-3}$, $r \neq 3$

7. $m+2$, $m \neq -1$

8. $\frac{1}{a-3}$, $a \neq 2$, $a \neq 3$

9. $\frac{3}{b+2}$, $b \neq -2$

10. $\frac{r}{r+1}$, $r \neq 1$, $r \neq -1$

11. $\frac{x+1}{x}$, $x \neq 0$, $x \neq 1$

12. $\frac{x+3}{x-4}$, $x \neq 2$, $x \neq 2$

13. $x - 2$

Lesson 11.4

1. $2t$, $t \neq 0$, $t \neq 1$

2. $\frac{2}{3}$, $a \neq 0$, $a \neq \frac{3}{2}$

3. $\frac{c+3}{2}$, $c \neq -3$

4. $\frac{4}{d+5}$, $d \neq 0$, $d \neq -3$, $d \neq -5$

5. $2(x-3)$, $x \neq -3$

6. $y^2 - 1$, $y \neq -2$, $y \neq 2$

7. $\frac{13x}{20}$

8. $\frac{37}{6b}$, $b \neq 0$

9. $-\frac{7}{12t}$, $t \neq 0$

10. $\frac{5m}{2(m-1)}$, $m \neq 1$

11. $\frac{c}{2(c-4)}$, $c \neq 4$

12. $\frac{3y^2 + 2y}{(y+2)(y-2)}$, $y \neq -2$, $y \neq 2$

Lesson 11.5

1. -6

2. $-\frac{6}{13}$ and 2, $d \neq 0$

3. 3, $h \neq 0$, $h \neq 6$

4. -3, $x \neq 4$, $x \neq 2$

5. $2\frac{2}{5}$ hours, or 2 hours and 24 minutes

6. $7\frac{1}{2}$ hours, or 7 hours and 30 minutes

Lesson 11.6

1. $3x - 6 = 15$ Given
   $3x = 21$ Addition Property
   $x = 7$ Division Property

2. $8x^2 = 72$ Given
   $x^2 = 9$ Division Property
   $x = \pm 3$ Property of Square Roots

3. $x^2 - x = 12$ Given
   $x^2 - x - 12 = 0$ Subtraction Property
   $(x-4)(x+3) = 0$ Factoring
   $x = 4$ or $x = -3$ Zero-Product Property

4. $\frac{5}{x} = \frac{x-3}{2}$ Given
   $2(5) = x(x-3)$ Cross Products
   $10 = x^2 - 3x$ Distributive Property
   $x^2 - 3x - 10 = 0$ Subtraction Property
   $(x-5)(x+2) = 0$ Factoring
   $x = 5$ or $x = -2$ Zero-Product Property

5. Let $2x + 1$ represent one odd number.   Given
   Let $2y + 1$ represent another odd number.   Given
   $(2x + 1) + (2y + 1)$ Simplify
   $= 2x + 2y + 2$ Expressions
   $= 2(x + y + 1)$ Distributive Property
   The integer represented by $2(x + y + 1)$ is even because any integer in the form $2m$ is even.

Algebra 1   Answers

ANSWERS

6. Let $2x$ represent an even number. — Given
Let $2x + 1$ represent the next odd number — One more than an even number is an odd number.
$(2x)(2x + 1) = 4x^2 + 2x$ — Distributive Property
$\qquad = 2(2x^2 + x)$ — Distributive Property
The integer represented by $2(2x^2 + x)$ is even because any integer multiplied by 2 is even.

7. Let x represent an integer. — Given
Let $x + 1$ represent the next integer. — The next integer is 1 more.
$x + (x + 1) = 2x + 1$ — Distributive Property
The integer represented by $2x + 1$ is odd because any integer in the form $2x + 1$ is odd.

8. Let x represent any integer. — Given
Let y represent any integer. — Given
$3x$ and $3y$ are multiples of 3. — Any number multiplied by 3 is a multiple of 3.
$3x + 3y = 3(x + y)$ — Distributive Property

The integer represented by $3(x + y)$ is a multiple of 3 because any number multiplied by 3 is a multiple of 3.

Reteach—Chapter 12

Lesson 12.1

1. $4\sqrt{3}$

2. $10b\sqrt{2}$

3. $\dfrac{\sqrt{2c}}{c}, c \neq 0$

4. $\dfrac{a\sqrt{15}}{5}$

5. $7\sqrt{6}$

6. $6\sqrt{3}$

7. $12\sqrt{2}$

8. $6\sqrt{2}$

9. $-4\sqrt{5}$

10. $4\sqrt{2}$

11. $31\sqrt{3}$

12. $-\sqrt{6} - 6$

13. $2\sqrt{3} + 5\sqrt{2}$

14. $5\sqrt{3} - 6$

15. $3\sqrt{2} + 2\sqrt{3}$

16. -1

17. $12 - 4\sqrt{2} + 6\sqrt{3} - 2\sqrt{6}$

18. $8 + 2\sqrt{15}$

Lesson 12.2

1. $x = 10$

2. $x = 11$

3. $x = 23$

4. $x = 8$

5. $x = 5$

6. $x = 2$

7. $x = 4\sqrt{15}$ and $x = -4\sqrt{15}$

8. $x = 25$

9. $x = 2$

10. $x = 2$

11. $x = 3$ and $x = -5$

12. $x = -3 + 2\sqrt{5}$ and $x = -3 - 2\sqrt{5}$

13. $x = 12$

ANSWERS

14. $x = 2$ and $x = 0$

15. $x = 1$

Lesson 12.3

1. 5
2. 12
3. 15
4. about 8.49 inches
5. about 8.7 meters
6. 10 miles
7. 75 miles
8. about 27.73 meters
9. about 10.61 feet

Lesson 12.4

1. 10 units
2. 9 units
3. $\sqrt{65}$ units, or about 8.06 units
4. length of $\overline{LM}$: $\sqrt{50}$, length of $\overline{MN}$: $\sqrt{18}$; length of $\overline{LN}$: $\sqrt{68}$.

   Because $\left(\sqrt{50}\right)^2 + \left(\sqrt{18}\right)^2 = \left(\sqrt{68}\right)^2$, $\triangle LMN$ is a right triangle.
5. $M\left(-\frac{7}{2}, 2\right)$
6. $M\left(-3, -\frac{5}{2}\right)$
7. $M(-2, -4)$
8. $E(10, 4)$
9. $C(1, -1)$

Lesson 12.5

1. $x^2 + y^2 = 25$
2. $x^2 + y^2 = 6.25$
3. $x^2 + y^2 = \frac{9}{16}$
4. $C(-5, -1); r = 6$
5. $C(4, 1); r = 3\sqrt{3}$
6. $C(3, 3); r = 2.8$
7. $\overline{BC} = 3\sqrt{5}, \overline{CA} = 3\sqrt{5}$
8. $\overline{AC} = 2\sqrt{13}, \overline{BD} = 2\sqrt{13}$
9. $\overline{AC} = \sqrt{(0-a)^2 + (0-a)^2} = \sqrt{a^2 + a^2} = \sqrt{2a^2} = a\sqrt{2}$
   $\overline{BD} = \sqrt{(0-a)^2 + (0-a)^2} = \sqrt{a^2 + a^2} = \sqrt{2a^2} = a\sqrt{2}$

Lesson 12.6

1. 1.333
2. 0.577
3. 1.875
4. 0.364
5. 1
6. 3.078
7. 9.4 feet
8. 23.3 meters
9. 10.4 feet

Lesson 12.7

1. $\sin A = 0.6; \cos A = 0.8$
2. $\sin A \approx 0.471; \cos A \approx 0.882$
3. $\sin A \approx 0.667; \cos A \approx 0.745$

Algebra 1   Answers   **203**

ANSWERS

4. sin 25° ≈ 0.423; cos 25° ≈ 0.906

5. sin 45° ≈ 0.707; cos 45° ≈ 0.707

6. sin 90° = 1; cos 90° = 0

7. 28.7 feet

8. 388.2 meters

9. 319.3 feet

Lesson 12.8

1. $\begin{bmatrix} 4.7 & -7.9 & 15 \\ -0.8 & 0 & 1 \end{bmatrix}$

2. Not possible; the matrices do not have the same dimensions.

3. $\begin{bmatrix} 2.5 & 3.5 \\ 0.5 & 7.5 \\ 7 & -5 \end{bmatrix}$

4. $\begin{bmatrix} -1.7 & 2.9 & -1 \\ -6.2 & -3.0 & 7 \end{bmatrix}$

5. Not possible; the matrices do not have the same dimensions.

6. $\begin{bmatrix} -5.5 & 3.5 \\ 4.5 & -5.0 \\ 7 & -3 \end{bmatrix}$

7. $\begin{bmatrix} 3 & -5 & 14 \\ -7 & -3 & -6 \end{bmatrix}$

8. Not possible; the matrices do not have the same dimensions.

9. $\begin{bmatrix} -36 & 36 \\ -144 & 144 \end{bmatrix}$

10. $\begin{bmatrix} -9 & -9 & 3 \\ -21 & 4 & 42 \end{bmatrix}$

Reteaching—Chapter 13

Lesson 13.1

1. H1, H2, H3, H4, H5, H6, T1, T2, T3, T4, T5, and T6

2. H1, H2, H3, H4, H5, H6, and T2

3. H4

4. H2, H4, H6, T1, T2, T3, T4, T5, and T6

5. T2, T4, and T6

6. $\frac{1}{4}$

7. $\frac{7}{11}$

8. $\frac{16}{101}$

Lesson 13.2

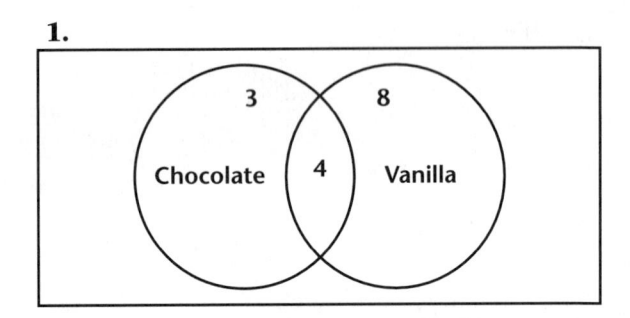

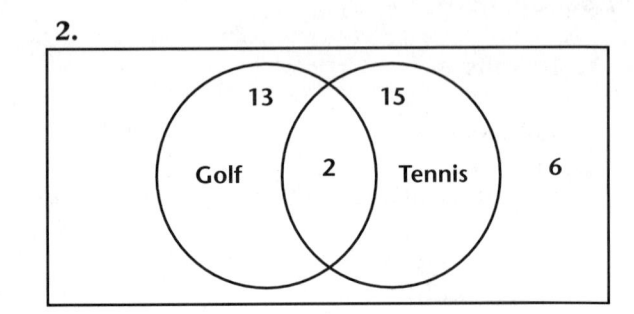

1.

```
┌─────────────────────────────────┐
│       ___        ___            │
│      /   \3    8/   \           │
│     │Choc.│  4  │Van.│          │
│      \___/      \___/           │
└─────────────────────────────────┘
```

2.

```
┌─────────────────────────────────┐
│      ___13    15___             │
│     /   \      /   \            │
│    │Golf │ 2  │Tennis│  6       │
│     \___/      \___/            │
└─────────────────────────────────┘
```

3. 135

4. 126

5. 91

6. 170

204 Answers Algebra 1

ANSWERS

Lesson 13.3

1. 60

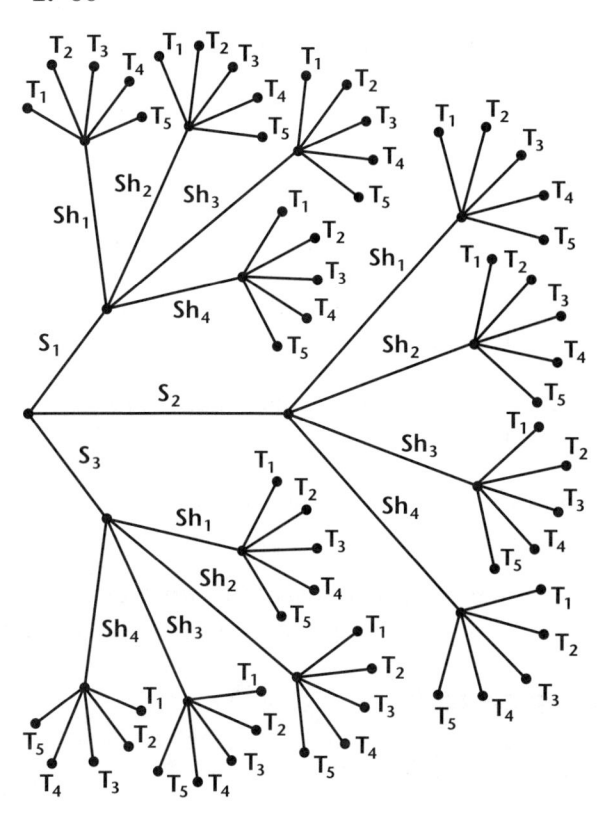

2. 36

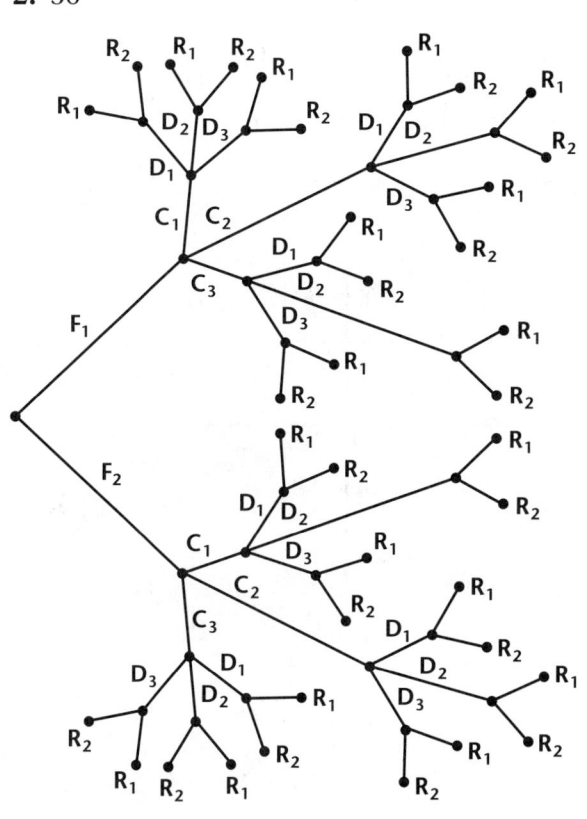

3. 6760

4. 100

5. 12

Lesson 13.4

1. no
2. yes
3. yes
4. yes
5. $\frac{1}{4}$
6. $\frac{1}{16}$
7. $\frac{1}{12}$
8. $\frac{1}{100}$

Lesson 13.5

1. Answers may vary. Sample answer: Let heads represent true and let tails represent false. Let each toss represent one trial. Generate 34 tosses (trials). The number of tails generated represents the number of students that guessed correctly.

2. Answers may vary. Sample answer: Let heads represent an even number and let tails represent an odd number. Let each toss represent one trial. Toss 52 times. The number of heads generated divided by 52 represents the probability that she starts with sit-ups.

3. Answers may vary. Sample answer: Use a coin as a random generator. Let heads represent red and let tails represent black. Each toss of the coin represents one trail. Generate 20 trials and record the results.

Algebra 1 — Answers 205

ANSWERS

4. Answers may vary. Sample answer: Use a number cube as a random generator. Let 1 represent red, 2 represent green, 3 represent yellow, 4 represent blue, 5 represent tan, and 6 represent dark brown. Each roll of the cube represents one trial. Generate 30 trials and record the results.

5. Answers may vary. Sample answer: Use a number cube and a coin as random generators. Let 1 or 2 represent rain on Saturday and let 3, 4, or 5 represent no rain. A roll of 6 means to roll again. A coin could be flipped for Sunday's result, using heads for rain and tails for no rain. Each pair of results would represent one trial. Generate 20 trials and record the results.

Reteaching—Chapter 14

Lesson 14.1

1. function
2. not a function
3. function
4. function
5. 38
6. −8
7. −15
8. $\frac{1}{9}$
9. not a function
10. function

Lesson 14.2

1. (−1, 4)
2. (9, 9)
3. (−10, 11)
4. (3, 2)

5. (13, 23)
6. (−17, 9)
7. $y = 2^x$; vertical translation 3 units up
8. $y = \frac{1}{x}$; vertical translation 2 units down
9. $y = x^2$; vertical translation 2 units up
10. $y = |x|$; vertical translation 4 units down
11. $y = 10^x$; horizontal translation 2 units to the right
12. $y = |x|$; horizontal translation 3 units to the left
13. $y = x^2$; horizontal translation 4 units to the right
14. $y = x^2$; vertical translation 3 units down

Lesson 14.3

1. $y = \frac{1}{x}$; scale factor of 3
2. $y = x^2$; scale factor of $\frac{1}{5}$
3. $y = |x|$; scale factor of 3
4. $y = 10^x$; scale factor of 2
5. The graph is stretched vertically by 3; it rises more sharply.

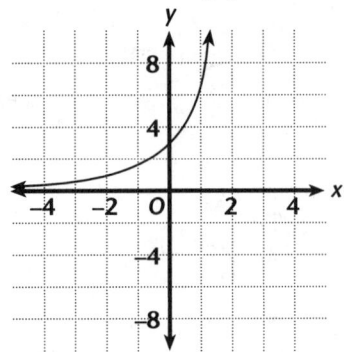

206 Answers Algebra 1

ANSWERS

6. The graph is stretched vertically by $\frac{1}{4}$; it is more spread out.

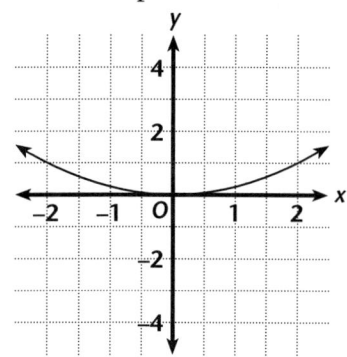

7. 2; stretched by a factor of 2

8. 3; stretched by a factor of 3

9. 5; stretched by a factor of 5

10. $\frac{1}{3}$; stretched by a factor of $\frac{1}{3}$

11. $\frac{1}{4}$; stretched by a factor of $\frac{1}{4}$

12. $\frac{1}{5}$; stretched by a factor of $\frac{1}{5}$

13. $\frac{2}{5}$; stretched by a factor of $\frac{2}{5}$

14. 0.65; stretched by a factor of 0.65

Lesson 14.4

1. yes
2. no
3. no
4. no
5. yes
6. yes
7. Sample answer: Choose (−2, 1), (0, −1), and (1, 0) from the original graph. Replace the second coordinate with its opposite. Graph (−2, −1), (0, 1), and (1, 0) to determine the reflection.

8. Sample answer: Choose (−2, 4), (−1, 2), and (0, 1) from the original graph. Replace the second coordinate with its opposite. Graph (−2, 4), (0, −1), and (0, −1) to determine the reflection.

Lesson 14.5

1. $y = |x|$
2. $y = x^2$
3. $y = 2^x$
4. $y = x$
5. $y = \frac{1}{x}$
6. $y = \frac{1}{x}$
7. $y = x^2$
8. $y = x^2$
9. The parent function has been shifted 2 units to the left, stretched by a factor of 3, reflected through the x-axis, and shifted 3 units down.
10. The parent function has been shifted 3 units to the right, stretched by a factor of 2, and shifted 1 unit up.
11. The parent function has been shifted 1 unit to the right, stretched by a factor of 4, and shifted 2 units up.
12. The parent function has stretched by a factor of 4 and shifted 2 units down.